ANALYTICAL CHEMISTRY SYMPOSIA SERIES - volume 5

biochemical and biological applications of isotachophoresis

ANALYTICAL CHEMISTRY SYMPOSIA SERIES

ANALYTICAL CHEMISTRY SYMPOSIA SERIES - volume 5

biochemical and biological applications of isotachophoresis

Proceedings of the First International Symposium, Baconfoy, May 4—5, 1979

edited by

Albert Adam

Laboratoire de Biologie Clinique, Centre Hospitalier de Sainte-Ode, B-6970 Baconfoy,
and Laboratoire de Radioimmunologie, C.H.U. Tour de Pathologie, Batiment B 23,
Université de Liège, Sart Tilman, B-4000 Liège, Belgium

and

Carlo Schots

LKB Instrument NV SA, Europalaan 23, B-9820 Ghent, Belgium

ELSEVIER SCIENTIFIC PUBLISHING COMPANY
Amsterdam — Oxford — New York   1980

ELSEVIER SCIENTIFIC PUBLISHING COMPANY
335 Jan van Galenstraat
P.O. Box 211, 1000 AE Amsterdam, The Netherlands

Distributors for the United States and Canada:

ELSEVIER/NORTH-HOLLAND INC.
52, Vanderbilt Avenue
New York, N.Y. 10017

ISBN 0-444-41891-1 (Vol. 5)
ISBN 0-444-41786-9 (Series)

Printed in The Netherlands

CONTENTS

Preface

The communications published in this book were presented at the
plenary session and at the poster session of the first international
symposium on "Biochemical and Biological Applications of Isotachophoresis" organized at the CENTRE HOSPITALIER DE SAINTE ODE (Belgium) on
the 4th and 5th of May 1979 in collaboration with the FONDATION
UNIVERSITAIRE LUXEMBOURGEOISE (FUL). Mr R. NACHEZ, National and
International President of war prisoners (F.N.A.P.G. and C.I.A.P.G.)
was honorary President of this symposium.

We would like to thank the people who contributed to the succes of
this symposium : the board of directors and the staff of the CENTRE
HOSPITALIER DE SAINTE ODE, and the FEDERATION DU TOURISME LUXEMBOUR-
GEOIS (F.T.L.B.).

We would particularly like to thank the FUL which makes the practical
realisation of these proceedings possible.

A. ADAM
C. SCHOTS
Organisers of the
Symposium.

Sainte Ode
January 1980

A. Adam and C. Schots (Editors), *Biochemical and Biological Applications of Isotachophoresis*
© 1980 Elsevier Scientific Publishing Company, Amsterdam — Printed in The Netherlands

ISOTACHOPHORESIS

A GENERAL INTRODUCTION

F.M. Everaerts and F.E.P. Mikkers

Department of Instrumental Analysis, Eindhoven   University of
Technology, Eindhoven, The Netherlands

1. HISTORICAL

The experiments of Lodge (ref. 1) and Whetman (refs. 2 and 3) were
the basis on which Kohlrausch (ref. 4) developed a theory of ionic
migration. The principles of electrophoresis can still be described
with the equation that he derived, including zone electrophoresis,
moving boundary electrophoresis and isotachophoresis.

It was not until 1923 that experiments of electrophoresis, other
than zone electrophoresis, were described. Kendall and Crittenden
(ref. 5) succeeded in separating rare earth metals and some simple
acids by, as they called it, the "ionic migrating method", which was
in fact isotachophoresis. The "moving-boundary" method of MacInnes
and Longsworth (ref. 6), which was used for the determination of
transport numbers was also in fact isotachophoresis. Then in 1942
Martin (ref. 7) separated chloride, acetate, aspartate and glutamate
by isotachophoresis, or displacement electrophoresis as he called it.

In the early sixties, simultaneously and independently, Konstantinov
and Oshkurova (ref. 8), and Everaerts (ref. 9) published results of
their isotachophoretic equipment, making use of an optic and a thermometric detector respectively.

Ornstein (ref. 10) and Davis (ref. 11) introduced disc electrophoresis, making use of a discontinuous electrolyte and gel system. The
first stage of this separation procedure is comparable with isotachophoresis. Schumacher and Studer (ref. 12) used isotachophoresis for
the separation of low-molecular-weight substances. Vestermark (ref. 13)
introduced a new term, "Cons. electrophoresis". Preetz (ref. 14) and
Preetz and Pfeifer (ref. 15) described the use of counterflow of

electrolyte in an isotachophoretic analysis.

Since 1970 an increasing number of scientists have contributed to
the development of isotachophoresis as can be seen from Acta Isota-
chophoretica.

2. THEORY

2.1 General equations

In isotachophoresis a steady-state configuration is obtained as
the result of a separation process that proceeds according to the
moving boundary principle. Although this separation process is a
transient state, it is governed by the same regulating function concept
as the steady state. A quantitative and qualitative description of transient state provides information on the time needed for an isotachophoretic separation (refs. 17 and 18). Moreover, such a description requires the definition of resolution and separability in isotachophoresis
and shows the results that can be expected from optimization procedures.

In electrophoresis the migration velocity, v, of a constituent i
is given by the product of effective mobility $\bar{m}_i$ and the local electrical field strength, E:

$$v_i = \bar{m}_i E \tag{1}$$

The electrical field strength is vectorial so the effective mobilities
can be taken as signed quantities, positive for constituents that migrate in a cathodic direction and negative for those migrating anodically.
As a constituent may consist of several forms of subspecies in rapid
equilibrium, the effective mobility represents an average ensemble.
Not dealing with constituents consisting of both positively and negatively charged subspecies in equilibrium, we can take concentrations
with a sign corresponding to the charge of the subspecies. Thus the
total constituent concentration, $\bar{c}_i$, is given by the sum of all of
the subspecies concentrations, c_n:

$$\bar{c}_i = \sum_n c_n \tag{2}$$

Following the mobility concept of Tiselius, the effective mobility is
given by:

$$\bar{m}_i = \sum_n \frac{c_n m_n}{\bar{c}_i} \tag{3}$$

where m_n is the ionic mobility of the subspecies. In dissociation equilibria the effective mobility can be evaluated using the degree of dissociation, α:

$$\bar{m}_i = \sum_n a_n m_n \tag{4}$$

The degree of dissociation can be calculated once the equilibrium constant, K, for the subspecies and the pH of the solution are known. For a restricted pH range a very useful relationship has been given by Hasselbalch (ref. 19):

$$pH = pK \pm log \left(\frac{1}{a} - 1 \right) \tag{5}$$

where pK is the negative logarithm of the protolysis constant; the positive sign holds for cationic subspecies and the negative sign for anionic subspecies.

All electrophoretic processes are essentially charge-transport processes that Obey Ohm's law. In electrophoresis this law is most conveniently expressed in terms of electrical current density, J, specific conductance, κ, and electrical field strength:

$$J = \kappa E \tag{6}$$

The specific conductance is given by the contributions of the individual constituents:

$$\kappa = F \sum_i \bar{c}_i \bar{m}_i \tag{7}$$

where F is the Faraday constant.

The equation of continuity states that for any electrophoretic process (ref. 6)

$$\frac{\partial}{\partial t} \cdot \bar{c}_i = - \frac{\partial}{\partial x} \left(\frac{\partial}{\partial x} D_i \bar{c}_i - v_i \bar{c}_i \right) \tag{8}$$

where t and x are time and place coordinates, respectively, and D is the diffusion coefficient. Neglecting diffusional dispersion we can apply eqn. 8 for each constituent and the overall summation of the

4

constituents gives

$$\frac{\partial}{\partial t} \sum_i \bar{c}_i = \frac{\partial}{\partial x} \cdot E \sum_i \bar{m}_i c_i \qquad (9)$$

In combination with the specific conductance (eqn. 7) and modified Ohm's law (eqn. 6), it follows that

$$\frac{\partial}{\partial t} \sum_i \bar{c}_i = 0 \ or \ \sum_i \bar{c}_i = constant \qquad (10)$$

For monovalent weakly ionic constituents, eqn. 8.can be written as

$$\frac{\partial}{\partial t} \cdot \bar{c}_i = \frac{\partial}{\partial x} \cdot E m_i c_i \qquad (11)$$

where m_i and c_i are the mobility and the concentration, respectively, of the charged species i. Division by m_i and application of the resulting relationship for each constituent and overall summation gives

$$\frac{\partial}{\partial t} \sum_i \frac{\bar{c}_i}{m_i} = \frac{\partial}{\partial x} \cdot E \sum_i c_i \qquad (12)$$

Electroneutrality, however, demands that $\sum_i c_i = 0$, thus

$$\frac{\partial}{\partial t} \sum_i \frac{\bar{c}_i}{m_i} \equiv 0 \ or \ \sum_i \frac{\bar{c}_i}{m_i} = constant \qquad (13)$$

Eqn. 13 is well known as the Kohlrausch regulating function[4].
In an electrophoretic system different zones can be present, where a zone is defined (ref. 20) as a homogeneous solution demarcated by moving and/or stationary boundaries. We can apply the continuity principle (eqn. 8) to a boundary and derive the general form of the moving boundary equation

$$\bar{m}_i^K \bar{c}_i^K E^K - \bar{m}_i^{K+1} \bar{c}_i^{K+1} E^{K+1} = v^{K/K+1} \left(\bar{c}_i^K - \bar{c}_i^{K+1} \right) \qquad (14)$$

where $v^{K/K+1}$ represents the drift velocity of the separating boundary between the zones K and $K + 1$.
In the case of a stationary boundary, the boundary velocity is zero and eqn. 14 reduces to

$$\frac{m_i^{-K+1} c_i^{-K+1}}{m_i^{-K} c_i^{-K}} = \frac{E^K}{E^{K+1}} = constant \qquad (15)$$

From eqn. 15 it follows directly that for monovalent weak and strong electrolytes all ionic subspecies are diluted or concentrated over a stationary boundary to the same extent, because

$$\frac{c_i^{K+1}}{c_i^K} = constant \qquad (16)$$

In isotachophoresis sample constituents migrate in a stacked configuration, steady state, between a leading ionic constituent of high effective mobility and a terminating constituent of low effective mobility. From the moving boundary equation (eqn. 9) it follows directly that, in a separation compartment of uniform dimensions at constant electrical driving current, all boundary velocities within the isotachophoretic framework are equal and constant. According to Joule's law, heat generation will occur, resulting in different regimes that are moving or stationary.

In order to reduce the effects of temperature, relative mobilities, r, can be introduced. Obviously the leading constituent, L, provides the best reference mobility:

$$r_i = \frac{m_i}{m_L} \qquad (17)$$

Moreover, as in most isotachophoretic separations, only one counter constituent, C, will be present, the reduced mobility, k, can be introduced:

$$k_i = \frac{1 - r_C}{r_i - r_C} \qquad (18)$$

Using the derived equations it is possible to calculate all dynamic parameters of analytical importance. Moreover, model considerations can be extended to moving boundary electrophoresis as well as zone electrophoresis.

6

2.2 The criterion of separation

As in all differential migration methods, the criterion for separation in isotachophoresis depends simply on the fact that two ionogenic constituents will separate whenever their migration rates in the mixed state are different. For two constituents i and j, this means that according to eqn. 1 their effective mobilities in the mixed state must be different:

$$\frac{\bar{m}_i}{\bar{m}_j} \neq 1 \qquad (19)$$

When the effective mobility of i is higher than that of j the latter constituent will migrate behind the former. Consequently, two monovalent weakly anionic constituents will fail to separate when the pH of the mixed state, pH^{mo}, is given by

$$pH^m = pH^{mo} = pK_j + log\left(\frac{1 - \dfrac{r_j K_j}{r_i K_i}}{\dfrac{r_j}{r_i} - 1}\right) \qquad (20)$$

where K_i and K_j are the protolysis constants for the subspecies of the constituents i and j. When the more mobile constituent has a higher protolysis constant, we are dealing with a "straight" pair of constituents; when the more mobile constituent has a lower protolysis constant, we have a "reserved" pair of constituents, for which the separation configuration is a function of the pH.

2.3 Resolution

Once the criterion for separation has been satisfied, the time needed for resolution becomes important. When a constituent zone contains all of the sampled amount, resolution has been obtained for that constituent. We therefore define the resolution, R, as the separated fractional amount of the constituent:

$$R_i = \frac{separated\ amount\ of\ i}{sampled\ amount\ of\ i} \qquad (21)$$

From this definition it follows that during the separation process the resolution increases from zero to its maxima value, unity. Constituents that fail to separate remain at zero resolution and can be termed ideally mixed zones (ref. 21).

Complete separation of a sample requires the resolution values of all constituents of interest to be unity. Maxima speed of separation is obtained whenever the resolution rate, $\partial/\partial t . R_i$, is optimized during the separation process. As expected, the resolution and its time derivatives are complex functions of the constituents involved and the driving forces applied. Moreover, the mathematical intricacy involved in calculating optimal process variables increases rapidly with increasing number and complexity of the sample constituents. For strong electrolytes relevant mathematical formulations have been published (refs. 22 and 23) but most separations nowadays concern weak electrolytes. In this case dissociation equilibria, and therefore a a proper choice of pH, are tools in the control and optimization of the separation process. When dealing with complex formation, association equilibria should be involved. Others (ref. 24) have suggested that the difference in migration rates, $v_i - v_j$, is of decisive importance in separation. However, in isotachophoresis and moving boundary electrophoresis this does not apply, and in these instances it is more benefical to optimize the ratio of the migration rates; for example, v_i/v_j. Whereas the velocity difference will reach a maximal value as a function of pH (ref. 24), the ratio shows no such optimum (ref. 18). As the local electrical field strength for both constituents will be the same, it follows directly that eqn. 19 must be maximized or minimized, depending on the migration configuration. On introducing equilibrium constants and ionic mobilities it follows that in anionic separations the lowest pH will give the better mobility ratio, and *vice versa* for cationic separations. It should be emphasized, however, that pH extremes have only limited experimental applicability and that practical considerations often govern the proper choice of pH. Moreover, a low numerical value of the effective mobility will induce a high electrical field strength to obtain an appreciable migration rate and other electrokinetic effects may then prevail.

2.4 Steady state

A unique feature of isotachophoresis is that, once the separation process has been completed, all electrophoretic parameters remain

constant with time. Assuming a uniform current density, all sample
constituents within the leading-terminating framework will migrate
at identical speeds. Moreover, at constant current density local
migration rates will be constant. In this steady state, resolution
values of stacked constituents will be either unity or zero.

The basic features of steady-state configurations have been extensively discussed.(ref. 16).

2.5 The separation process

The applicability of the above equations and definitions and the
resulting implications are best illustrated by using a relatively
simple two-component sample. We shall deal with the case where all
constituents involved are monovalent weak electrolytes. Although
essentially immaterial, we shall consider a separation compartment
of uniform dimensions at a constant electrical driving current and a
constant temperature. The separation process and some relevant information are given in Fig. 3.

It should be emphasized that within the separator three different
regions are present and each has its own regulating behaviour. The
regulating functions (eqns. 10 and 13) are the mathematical expression
of this regulating behaviour and locally they cannot be overruled by
the electrophoretic process (ref. 4). All changes in electrophoretic
parameters, such as concentration, pH and conductance, will be in
agreement with the local regulating function. Applying eqn. 16 to the
stationary boundary between the separation and the sampling compartment
it follows that

$$\phi = \frac{c_B^{M*}}{c_A^{M*}} = \frac{c_B^M}{c_A^M} = constant \qquad (22)$$

Hence, the sampling ratio, ϕ, for the charged subspecies is constant.
Taking the leading electrolyte as frame of reference, the regulating
functions (eqns. 10 and 13) will result in

$$\bar{c}_L^L + \bar{c}_C^L = \bar{c}_A^M + \bar{c}_B^M + \bar{c}_C^M \tag{23}$$

and

$$\frac{\bar{c}_L^L}{r_L} + \frac{\bar{c}_C^L}{r_C} = \frac{\bar{c}_A^M}{r_A} + \frac{\bar{c}_B^M}{r_B} + \frac{\bar{c}_C^M}{r_C} \tag{24}$$

where C is the counter constituent common to all constituents to be separated. The equilibrium relationships (eqn. 5) and electroneutrality imply that

$$\bar{c}_A^M = c_A^M \left[1 + 10^{\pm(pK_A - pH^M)} \right] \tag{25}$$

$$\bar{c}_B^M = \phi c_A^M \left[1 + 10^{\pm(pK_B - pH^M)} \right]$$

$$\bar{c}_C^M = -(1 + \phi) c_A^M \left[1 + 10^{\pm(pH^M - pK_C)} \right]$$

Combining eqns. 23, 24 and 25 we obtain

$$\bar{c}_L^L + \bar{c}_C^L = c_A^M \left[\frac{1}{a_A^M} + \frac{\phi}{a_B^M} - \frac{(1 + \phi)}{a_C^M} \right] \tag{26}$$

and

$$c_L^L(1 - r_C^- = c_A^M \left[\frac{r_A - r_C}{r_A a_A^M} + \frac{\phi(r_B - r_C)}{r_B a_B^M} \right] \tag{27}$$

We now introduce the relative leading concentration

$$e = \frac{\bar{c}_C^L}{\bar{c}_L^L}$$

and the reduced mobility. Elimination of c_A^M gives a quadratic equation for the proton concentration in the mixed zone. Only one root will have physical significance:

$$a \cdot 10^{2pH^M} + b \cdot 10^{pH^M} + c = 0 \tag{28}$$

The constants for the equation are given in Table 1. Once the pH in the mixed zone has been calculated, all dynamic process variables can be calculated using eqns. 1-27. Moreover, steady-state configurations are obtained by the introduction of zero or infinite sampling ratios. Moving boundary experiments can be simulated by introduction of a high load of sample. Computerization allows multiple calculations of all dynamic process variables.

TABLE 1
Dynamic constants for the pH of the mixed zone (eqn. 28)

Anionic constituents	Cationic constituents
$a = 10^{-pK}C(1 + \phi)$	$a = 10^{-pK_A}\left(\dfrac{1+\rho}{r_A k_A} - 1\right) + \phi \cdot 10^{-pK_B}\left(\dfrac{1+\rho}{r_B k_B} - 1\right)$
$b = (1 + \rho)\left(\dfrac{1}{r_A k_A} + \dfrac{\phi}{r_B k_B}\right)$	$b = (1 + \rho)\left(\dfrac{1}{r_A k_A} + \dfrac{\phi}{r_B k_B}\right)$
$c = 10^{pK_A}\left(\dfrac{1+\rho}{r_A k_A} - 1\right) + \phi \cdot 10^{pK_B}\left(\dfrac{1+\rho}{r_B k_B} - 1\right)$	$c = 10^{pK}C(1 + \phi)$

2.6 The pH of the mixed zone

As the criterion for separation has to be satisfied and the ratio of effective constituent mobilities must be optimized, the pH of the mixed zone is of decisive importance. According to eqn. 28, this pH is influenced by the physicochemical characteristics of the species to be separated and the counter constituent, by the sampling ratio, ϕ, and the relative leading concentration, ρ. The latter parameter is closely related to the pH of the leading electrolyte and the former to the pH of the sample. We shall consider anionic separations, but equivalent relationships and conclusions can be made for cationic separations.

In isotachophoresis the leading constituent must have an effective mobility, so strong ionic species like chloride are commonly used (ref. 16). In this instance it follows that

$$- \infty < \rho = - \frac{1}{a_C^L} < -1 \tag{29}$$

At $\rho = -1$ the counter constituent is used far below its pK value and it behaves like a strongly ionic species. In this event the leading electrolyte has no buffering capacity. At $\rho = -2$ the counter constituent is used at its pK value, $pH^L = pKC$, and therefore it exhibits its full buffering capabilities. High negative values for the relative leading concentration again imply low buffering. Moreover, the concentration of the counter constituent will be high in comparison with that of the leading constituent, which can be favourable in complex formation. It is easily shown that for increasing $pH^L - pK_{i,j}$, i.e. the constituents to be separated are only partially dissociated at the pH of the leading electrolyte, $pH^M - pH^L$ will increase. Constituents that are completely ionized at the pH of the leading electrolyte will induce only a slight elevation of pH^M and therefore will be separated as strong electrolytes. Counter constituents with a low pK value in comparison with the pH of the leading electrolyte show a tendency to diminish this increase in pH^M. When the leading constituent is a strongly ionic species the pH of any following zone will be higher than the pH of the leading zone. If, however, a weak constituent is chosen as the leading constituent, negative pH steps can occur under appropriate conditions (ref. 16).

Time of resolution and length of resolution

Resolution has been defined as the separated fractional amount of the constituent under investigation. Maximal resolution, $R = 1$, is obtained whenever the constituent zone contains all of the sampled amount n. From Fig. 3 it can be concluded that the time for resolution of the constituent A can be expressed as a function of the boundary velocities $v_{L/A}$ and $v_{A/A+B}$

$$t_{res} = \frac{l_A}{v_{L/A} - v_{A/A+B}} \tag{30}$$

Using the appropriate relationships we obtain

$$
t_{res} = \frac{n_A F}{I} \left(\frac{1 + \phi . \dfrac{k_A}{k_B}}{1 - \dfrac{a_B^M r_B}{a_A^M r_A}} \right) \left(1 - \frac{r_C}{r_A} \right) \tag{31}
$$

Hence it follows that the time of resolution is a complex function of the concentration and the pH of both the leading electrolyte and the sample, of the sampled amount, the sampling ratio, the electric driving current and all ionic mobilities and dissociation constants involved. It should be noted that in eqn. 31 it is the ratio of the effective mobilities and not their difference that is important. Further, this equation emphasizes the importance of the pH of the mixed zone. For the length of separation compartment needed to contain the completely resolved state, x_{res}, it follows that

$$
x_{res} = \frac{n_A F}{0 c_L^{-L}} \left(\frac{1 + \phi . \dfrac{k_A}{k_B}}{1 - \dfrac{a_B^M r_B}{a_A^M r_A}} \right) \left(\frac{1}{k_A r_A} \right) = l_A \left(\frac{1 + \phi . \dfrac{k_A}{k_B}}{1 - \dfrac{a_B^M r_B}{a_A^M r_A}} \right) \tag{32}
$$

where O is the area of the separation compartment and l_A is the zone length of the resolved constituent A.

For a given sample and electrolyte system, the resolution length is independent of the applied current density or electrical field strength, whereas the time of resolution is inversely related to the electrical driving current. The load capacity of the column can be deduced from the resolution length. Obviously, a high load capacity is always favoured by a low resolution time, so we shall confine our considerations to the resolution time.
Taking limiting values for eqn. 11, it follows that

$$
\frac{n_A F}{I} < t_{res} < \infty \tag{33}
$$

The relationship between the sampled amount and the time of resolution is obviously linear. Moreover, for a two-constituent sample, resolution for both constituents will be obtained simultaneously. From eqns. 28 and 31 it can be concluded that both the time of resolution and the pH of the mixed zone are affected by the mobility of the common counter constituent.

As eqn. 31 is a function of the effective constituent mobilities in the mixed zone, the pH of the zone is very important. Recognizing that all mixed zone characteristics are determined by the leading electrolyte as well as by the sample, it is obvious that the relative leading concentration, ρ, and the sampling ratio, ϕ, can be used in optimization procedures. Both ϕ and ρ are functions of pH and can be chosen arbitrarily within practical limitations.

It should be noted that the effect on the time of resolution is appreciable. When the ionic mobilities are almost equal, it follows that a low pH^L must be chosen in order to obtain an acceptable resolution time. For species that already have large differences in their ionic mobilities, the effect of decreasing the pH of the leading electrolyte is less pronounced. Greater differences in equilibrium constants give even higher results.

2.8 Resolution of resolution rate

It follows directly from eqn. 31 that for a given sample and electrolyte system the amount of constituent A resolved into its proper zone is given by

$$n_A^{separated} = \frac{tI}{F} \left(\frac{1 - \dfrac{a_B^M r_B}{a_A^M r_A}}{1 + \psi \cdot \dfrac{k_A}{k_B}} \right) \left(1 - \frac{r_C}{r_A} \right)^{-1} \tag{34}$$

Thus, for the effective resolution $\bar{R}_A$ and its time derivative, the effective resolution rate, it follows that

$$\bar{R}_A = \frac{t}{t_{res}} \quad and \quad \frac{\partial}{\partial t} \cdot \bar{R}_A = \frac{1}{t_{res}} \tag{35}$$

It must be emphasized that, owing to eqn. 30, this resolution is an average. The actual resolution, according to eqn. 21, can be different from the effective resolution, because of discontinuities in the separation process. As a high resolution rate is always desirable, all conclusions regarding the time of resolution will apply. Eqn. 34, however, offers a unique possibility for deriving the dimensionless separation number S. Differentiation of the separated amount with respect to time, and multiplication by F/I , gives

$$S_A = \frac{F}{I} \cdot \frac{\partial}{\partial t} \cdot n_A \left(\frac{1 - \frac{a_B^M r_B}{a_A^M r_A}}{1 + \phi \cdot \frac{k_B}{k_A}} \right) \left(\frac{r_A}{r_A - r_C} \right) \tag{36}$$

The advantage of this separation number si that it is essentially independent of the amount of sample, column geometry and electrical driving current. The physical significance of the dimensionless separation number is that it gives the efficiency of the separation process. Taking limiting values, it follows that

$$0 < S_A < 1 \tag{37}$$

The separation number for the constituent B is closely related to that of A , as

$$S_B = \chi S_A \tag{38}$$

Where χ is the molar concentration ratio, $\bar{c}_B/\bar{c}_A$, in the sample The relationship between X, ϕ and the pH of the sample is straight-forward (ref. 13). From eqn. 36 it follows that the sampling ratio ϕ, may have a decisive influence on the separation number. Fig. 1 shows this influence for different ionic constituents. The counter constituent has again been chosen for its maximal buffering capacity in the leading electrolyte.

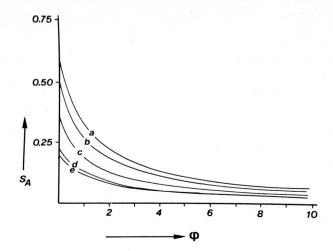

Fig. 1. Influence of the sampling ratio on the dimensionless separation
  number, ϕ = Variable, ρ = -2, r_A = 0.6, r_B = 0.4, r_C = -0.4
  (a) pK_A = 4, pK_B = 9, pK_C = 6; or pK_A = pK_B = 6 pK_C = 4.
  (b) pK_A = 4, pK_B = 7, pK_A = 6; or pK_A = pK_B = 5, pK_C = 4.
  (c) pK_A = 4, pK_B = 6, pK_C = 6; or pK_A = pK_B = 4, pK_C = 4.
  (d) pK_A = 4 pK_B = 5, pK_C = 6; or pK_A = pK_B = 5, pK_C = 6.
  (e) pK_A =.4, pK_B = 4, K_C = 6.

It follows that the separation number decreases rapidly with increase
in sampling ratio. By introducing limiting values, the transport
numbers (ref. 12) for the constituents in their resolved zones are
obtained: S_A = 0.6 and S_B = 0.5. Numerical calculations reveal that
many of the curves that are obtained when the physical parameters
pK and m are varied show congruent behaviour (Fig. 1). At low sampling
ratios a large difference in pK values will induce a high separation
number for the more mobile constituent. At high sampling ratios this
effect is much less pronounced. Moreover, from curves d and e in
Fig. 1 it follows that for the separation of a straight species pair
a low pH of the leading electrolyte is favourable. It should be
recognized that exact data for constituents are generally not known
and therefore an appreciable variation in the input data has to be
taken into consideration. Reasons for these variations are obvious:
lack of data unreliable data, temperature effects, activity effects,
etc. The broadening effect due to the parameter fluctuations, however,
is marginal, as many of these are counter-active.

16

2.9 Time of detection and load capacity

Eqn. 32 suggests that a fixed-point detector must be located at x_{res} from the sampling compartment. From Fig. 3, however, it follows that this is not always the case, as detection can already have commenced before the sample has been completely resolved. As the criterion for detection is only resolved constituents must be detected i.e, the mixed zone should resolve the moment it reaches the detection system. Hence, for the minimal length at which the detector must be located, x_{det}, it follows that

$$x_{det} = t_{res} V_{A/AB} \tag{39}$$

and, for the moment at which detection must be started, t_{det}

$$t_{det} = x_{det}/V_L \tag{40}$$

It follows directly that the time of resolution will be greater than or equal to the time of detection, as holds for the resolution distance and the detection distance. Using the appropriate relationships we obtain

$$\frac{x_{det}}{x_{res}} = \frac{t_{det}}{t_{res}} = \frac{\phi \cdot \dfrac{K_B}{K_A} + \dfrac{a_B^M r_B}{a_B^M r_A}}{1 + \phi \cdot \dfrac{K_B}{K_A}} \tag{41}$$

For a non-scanning detector it is important to minimize both x_{det} and t_{det}, and optimization procedures are analogous to the minimization of the resolution time. In common practice, however, the detector will be located at a fixed position in the separation compartment, x_{detfix}, so it is impossible to choose the actual length of the separation compartment. For the maximal load capacity, n^{max}, for the column we obtain

$$n_A^{max} = n_L^{load} r_A K_A \quad \cdot \quad \frac{1 - \dfrac{a_B^M r_B}{a_A^M r_A}}{\phi \cdot \dfrac{K_A}{K_B} + \dfrac{a_B^M r_B}{a_A^M r_A}} \tag{42}$$

where n_L^{load} is the amount of the leading constituent filling the
separation compartment from the sampling compartment to the detector.
The maximal load capacity for the second constituent follows directly
from the given definitions. Moreover, eqn. 42 can be transformed
directly into a time-based or distance-based form using

$$n_L^{load} = O x_{detfix} \bar{c}_L^L = t_{detfix} \cdot \frac{1}{F(1 - r_C)} \tag{43}$$

A maximal load capacity is obtained by minimizing the time for
resolution.

Experimentally, the amount of leading constituent can be determined
by injecting a known amount of leading constituent, Δn_L , and measuring
the resulting time delay, Δt, with respect to t_{detfix}:

$$n_L^{load} = t_{detfix} \cdot \frac{\Delta n_L}{\Delta t} \tag{44}$$

Moreover, because for a one-constituent zone the dimensionless
separation number, S, is identical to the transport number, T, the
experimental and theoretical transport efficiency can be compared:

$$T = \frac{F}{I} \cdot \frac{n_L^{load}}{t_{detfix}} = \frac{m_i}{\sum_i m_i} \tag{45}$$

where m_i is the ionic mobility of the constituent i. It should be noted
that the transport number of monovalent weakly ionic constituents
does not contain effective mobilities, but rather ionic mobilities,
because, owing to electroneutrality, the degrees of dissociation
cancels out.

2.10 Discussion and conclusion

In the transient-state model we neglected several secondary effects, for example, temperature distribution and activity effects. Although these effects are not always marginal, they will generally not imply other indications. With regard to uneven temperature distributions, either longitudinal or transverse, it should be emphasized that their effect will be deleterious only under extreme operating conditions. Working at moderate current densities, without excessive cooling, convection disturbances are negligible and temperature differences can be well controlled. In special cases, temperature effects can have a favourable influence on separation but so far temperature programming has not been studied. A fundamental question concerns the applicability of the transient-state model under extreme pH values of the electrolytes. The hydroxyl and/or proton concentration can be introduced into the specific conductance without difficulty. It has been suggested, however, that it is not necessary to incorporate the hydrogen constituent into the moving boundary equation. Neglecting the solvent effect at low or high pH gives differences in the zone charac-teristics compared with those of steady-state models. The differences, however, are small and their experimental significance is still under investigation.

The applicability of the transient-state model and the resulting implications have been considered for a sample containing two mono-valent weakly ionic constituents. The relative simplicity of the model allows a fundamental understanding of the isotachophoretic separation process and provides a realistic view of optimization procedures. It is obvious that the model could be extended to multivalent weak elec-trolytes. The efficiency of such considerations will be poor as no other indications will be found. Concerning multicomponent samples, is has already been mentioned that the mathematical intricacy increases rapidly with an increase in the number of constituents. For strong electrolytes extension of the model is not difficult but is laborious for practical purposes. The model given already indicates clearly the importance of physicochemical and operational parameters. For multicomponent samples optimization procedures will generally be difficult, as their success depends largely on the constituents involved.

Optimal separation has been identified with a resolution of unity of the constituent of interest. As the separation boundary between two resolved zones will always have a finite interfacial thickness, in which major concentration changes will occur, the ideal resolution of unity can never be obtained. On most occasions, however, when the sharpening effect of the applied electrical field and the dispersion by diffusion and convection forces are optimized, the interfacial thickness is so small that it cannot be detected in these instances. Therefore, deviations of the resolution from unity can be neglected. For very small zone lengths (<0.05 mm), the interfacial thickness has a deleterious effect on resolution. Moreover, the presence of a zone profile, parabolic or otherwise, makes reliable detection of very small zones very difficult, as not the actual zone length but the zone profile will be measured. Therefore, for accurate quantitative determinations the zone length should be reasonable.

The lower limit of resolution represents the case in which the constituent of interest does not separate at all and therefore forms a mixed zone with another constituent. In an ideal mixed zone a homogeneous concentration distribution should be present. Experimental steady-state mixed zones will have zero resolution but generally with a non uniform concentration distribution (ref. 21).

For a completely resolved sample, the resolution of all components should be unity. In practice, however, it is sufficient to obtain maximal resolution for the constituent of interest. It has been shown that for a two-constituent sample, the resolution will generally increase linearly with time. The exceptional behaviour of reversed pairs with respect to continuity and linearity has been indicated. With multi-component systems discontinuities and non-linear behaviour will be encountered more frequently.

A maximal resolution rate is obtained when the ratio of the effective mobilities of the constituents is minimized or maximized. Very low numerical values for the effective mobilities imply high electrical gradients. The resulting temperature effects and non-electrophoretic phenomena in this instance may have deleterious effects on resolution and resolution rate. Minimization or maximization procedures naturally must fit within the constraints of the leading-terminating electrolytes. It must be emphasized that the ratio of effective mobilities from the completely resolved (i.e. steady-state) configuration gives only indirect information about the separability and separation efficiency of two constituents. Hence constituents showing virtually no difference in steady-state effective mobilities can nevertheless sometimes be separated efficiently (ref. 18). In such instances the transient-state model shows that the pH of the mixed zone is the separation-determining parameter.

In the steady state, constituents will generally migrate in order of effective mobilities, i.e. the effective mobilities decrease from leading to terminating electrolyte. In special cases, however, a constituent with low effective mobility can migrate isotachophoretically in front of a constituent with a relatively high effective mobility. Such separation configurations have been called "enforced isotachophoresis" (ref. 16) and are stable with respect to time.

It follows that measurement of the step heights of single constituents gives only an indication of the separation configuration for a mixture of constituents. Moreover, it has been shown that, depending on pH, constituents can migrate in a different steady-state configuration. The importance of the pH of the leading electrolyte in this respect has been extensively discussed. It is generally assumed that the nature of the sample, especially its pH and concentration, has no influence on the steady state. The transient-state model, however, reveals the importance of the pH of the sample. Dealing with a reversed pair of sample constituents near the critical pH of the leading electrolyte, at which a reversal of order can occur, the pH of the sample may, theoretically, be the deciding factor. Hence it must be concluded that the steady-state characteristics of the zone are not influenced by the pH of the sample, but this (ref. 17) can affect the separation configuration. However, in practice this will not generally occur. It is obvious that, for constituents separating very slowly, it will be difficult to conclude whether the steady state has been reached or not. This holds especially for complex mixtures such as natural protein mixtures, in which the numerous constituents, each with possible microheterogeneity, may give rise to a continuous mobility spectrum. Such complex mixtures require a relatively long separation time. Obviously the use of spacers for such samples, whether ampholines or discrete substances, will decrease the efficiency of the separation process, but can increase the interpretability. Whenever possible the use of discrete spacers at low concentration is to be preferred.

The critical point of separability has been expressed in the criterion for separation; i.e. the ratio of effective mobilities in the mixed state should be different from unity. It should be recognized that this criterion gives only an academic answer to the question of whether constituents can be separated or not. Dealing with separability in its limiting case it is obvious that dispersion factors become important and should be incorporated into the equation of continuity and its resulting relationships. Relevant mathematical formulations have already been given for calculating the structure of separation boundaries in isotachophoresis.

Dispersion, however, may have several causes (for example, temperature distribution, osmotic and hydrodynamic flow and density gradients) and may exceed diffusional dispersion by several orders of magnitude. This overall dispersion is closely related to the chosen operating conditions and the design of the equipment. Allowance can be made for such dispersion factors, but the resulting uncertainty in the criterion for separation causes this to remain academic. The model presented clearly indicates that in dealing with practical separability other parameters are important, such as resolution rate, time for resolution, time of detection and load capacity. It has been shown that in addition to the physicochemical characteristics of the constituents, the sample load, the sampling ratio, the pH of the leading electrolyte and the sample and the applied electrical driving current determine in practice whether resolution can be obtained within an acceptable time. For separations in which a long analysis time is needed, ultrapure electrolyte systems must be used to prevent a progressive decay of the steady-state configuration. In optimization procedures three rationales can be recognized, which of course, are not completely independent:

(i) The electrical driving current effects directly the analysis time. As the time for resolution is inversely related to the electrical driving current, it is obvious that this operational parameter must be maximized. In practice this will mean that a compromise must be found between the quantitative and qualitative accuracy required and the allowable driving current. The electrical driving current, if temperature effects are neglected, has no influence on the efficiency of the separation process, so the length of resolution, the location of the detection system and the load capacity are all independent of it.

This is consistent with the fact that only the current-time integral is important (ref. 4). In order to separate a given sample a definite number of coulombs are necessary and the time interval in which this amount must be delivered is immaterial.

When performing isotachophoretic analyses, it is therefore not necessary to work at a constant electrical driving current. Using a fixed point detector, however, a constant electrical driving current greatly facilitates the interpretation of the isotachophoretograms obtained. Further, the operating conditions are more easily standarized and better controlled.

(ii) The efficiency of the current transport is directly influenced by the mobility of the common counter constituent. The favourable effect of a counter constituent with a low ionic mobility is directly reflected in the time for resolution, time for detection, separation number and load capacity. In practice, however, only few substances will satisfy all requirements: low mobility, low buffering capacity and no UV absorption.

(iii) The efficiency of the separation process is determined by the properties of the mixed zone. The transient-state model shows that these properties are also governed by the nature of the leading electrolyte as well as the nature of the sample. Considering the ratio of effective constituent mobilities in the mixed state, it follows that, owing to the limited numerical extension of ionic mobilities, pH or complex formation provides the best optimization parameter. In anionic separations a low pH of both the leading electrolyte and the sample will favour a high resolution rate and a high separation number. For cationic separations a high pH will be preferable. The presence of reversed pairs of constituents may complicate the optimization procedure. In general, it can be taken that the pH values of the leading electrolyte and the sample should not differ too much. For known species the critical pH values at which separation will not occur can easily be calculated and hence can be avoided.

3. PRACTICAL EVALUATION

All experiments were performed using the isotachophoresis equipment developed by Everaerts et al. (ref. 16). The separation compartment consisted of PTFE narrowbore tubing with an I.D. of 0.45, 0.2 or 0.15 mm, corresponding to an O.D. of 0.75, 0.4 and 0.3 mm, respectively.

The direct, constant electrical driving current was obtained from a
modified Brandenburg (Thornton Heath, Great Britain) high voltage
power supply. Potential gradient detectors, used in either the
potential gradient or the conductance mode, were used for the determination of transient-state and steady-state characteristics. All
chemicals used were of pro analysi grade or additionally purified by
conventional methods. Operational systems are summarized in Table II.
Theoretical calculations were performed with the computerized
transient-state model, and physicochemical data were taken form refs.
25-27.

TABLE II

Operational systems

Parameter	System N°				
	1	2	3	4	5
pH of leading electrolyte	3.60	4.03	4.60	5.04	6.02
Leading constituent	Cl^-	Cl^-	Cl^-	Cl^-	Cl^-
Concentration (M)	0.01	0.01	0.01	0.01	0.01
Counter constituent[+]	BALA	GABA	EACA	CREAT	HIST
Terminating constituent	C_2H_5COOH	C_2H_5COOH	C_2H_5COOH	C_2H_5COOH	MES[++]
Concentration (M)	0.005	0.005	0.005	0.005	0.005
Additive to leading electrolyte	0.05 % MOWIOL[+++]	0.05 % MOWIOL	0.05 % MOWIOL	0.05 % MOWIOL	0.05 % MOWIOL
Temperature	Ambient	Ambient	Ambient	Ambient	Ambient
Electrical driving current (A/cm^2):					
d_t = 0.45 mm	0.0503	0.0503	0.0503	0.0503	0.0503
d_t = 0.20 mm	0.0796	0.0796	0.0796	0.0796	0.0796
d_t = 0.15 mm	0.1415	0.1415	0.1415	0.1415	0.1415

+ BALA = β-alanine; GABA = γ-aminobutyric acid; EACA = ε-aminocaproic
acid; CREAT = creatinine; HIST = histidine.

++ MES = 2-(N-morpholino)ethanesulphonic acid.

+++ MOWIOL = polyvinyl alcohol, N-88, Hoechst, Frankfurt, G.F.R.

Transient-state characteristics can be easily obtained experimentally and several important parameters can be evaluated directly as most
are interrelated. For isotachophoretic analyses it is most convenient
to use a separation compartment of well-defined and constant volume
and to apply a constant electrical driving current. Using a fixed
point detector and a given operational electrolyte system, all characteristics can be evaluated by measurement of all electrical, gradient
and time events. Because, under these conditions, the amount of the

leading constituent filling the separation compartment, n_L^{load}, is constant, the first boundary that reaches the detector will always be registered after the same time interval, t_{detfix}. Some experimental results are given in Table III. Obviously, a high transport efficiency,

TABLE III

Characteristics of the leading electrolyte

Parameter	Value	
Leading constituent, chloride	m_{Cl^-}	$= -77 - 10^{-5}$ cm^2/V . sec
Concentration	c_L	$= -0.01$ M
Counter constituent, γ-aminobutyric acid	m_{GABA}	$= 30$. 10^{-5} cm^2/V · sec
	pK_{GABA}	$= 4.03$
Electrical driving current	I	$= 80$ μA
Diameter of separation compartment	d_t	$= 0.45$ mm
Appearance of the first boundary	t_{detfix}	$= 1112$ sec
Amount of leading constituent samples	Δn_L	$= 100$ nmole
Time delay to amount sampled	Δt	$= 59.2$ sec
Response	$\dfrac{\Delta n_L}{\Delta t}$	$= 0.592$ nmole/sec
Load of leading constituent	n_L^{load}	$= 658$ nmole
Transport number: experimental	T_{exp}	$= 0.714$
theoretical	T_{theor}	$= 0.720$
Transport efficiency: experimental	ε_{exp}	$= 71$ %
theoretical	ε_{theor}	$= 72$ %

due to a low ionic mobility of the counter constituent, is always favourable as it guarantees efficient use of the power applied. For the operational systems given in Table II we can expect transport efficiencies of between 70 and 75 %. The characteristics of a separation process can be evaluated by the injection of known amounts of sample. An example of such a procedure is given in Fig. 2 and relevant parameters are summarized in Table IV. The fact that, at a constant load of leading constituent, the first boundary will always be detected at the same time interval, t_{detfix}, is illustrated in Fig. 2 by the resolution line L/A. The low coefficient of variation confirms the excellent performance of the equipment. Injection of a small amount of sample will cause two zones, stacked between the leading constituent L and the terminating constituent T . A sample load of

25

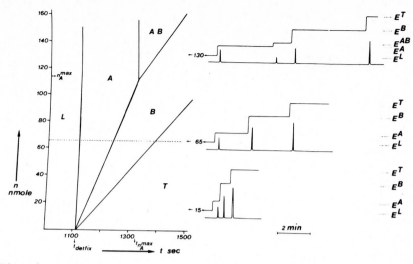

Fig. 2. Resolution for a two-constituent mixture. Operational system: Table II, system N° 2, and Table III. L = chloride; A = formic acid; B = glycolic acid; T = propionic acid; $E^{L,A,B,T}$ = electrical field strength; n = amount sampled. Sample: $\overline{C}^{sample}_{formate}$ = -0.005 M; $\overline{C}^{sample}_{glycolate}$ = -0.05 M; pH^{sample} = 3.00.

TABLE IV

Resolution data

For operational system see Fig. 2 and Table III. Resolution line: $n = at$ (nmole)

Boundary	N° of determinations	a		b		Coefficient of variation or Correlation coefficient
		Experimental	Theoretical	Experimental	Theoretical	
L/A	53	0	0	-1112	-1112	0.8 %
A/B	13	0.525	0.530	584	590	1.000
B/T	45	0.242	0.251	270	279	1.000
A/AB	6	0	0	-1328	-1317	0.4 %
AB/B	6	0.321	0.316	314	312	0.998

Parameter	Experimental	Theoretical
Load of leading constituent (n_L^{load})	658	647
Maximal sample load (n_A^{max})	113	108
Separation number (S_A)	0.103	0.099
(S_B)	0.103	0.099
Load capacity (C_{load})	0.172	0.167
Separation efficiency (ε, %)	21	20

1.3 µl of the constituent mixture (Fig. 2), where $n_A = 65$, will give
a time-based zone length of 124.5 sec for constituent A and detection
must be started 1112 sec after injection. The zone length of the
second constituent, B, will be 148.1 sec. Other sample loads give
proportional zone lengths. The characteristics of these steady-state
zones have already been discussed extensively and the close agreement
of the calculated and experimental resolution lines, $L/A, A/B$ and B/T,
indicates the reliability of the calculations.

As the separation compartment has a limited load capacity, at a
high load a mixed zone will be detected. The characteristics of these
mixed zones are determined by both the leading electrolyte and the
sample and are constant with time, as long as they exist. The time
interval, t_{nA}^{max}, at which the mixed zone will be detected, is again
constant, as illustrated in Fig. 2 by the resolution line A/AB.

$$t_{nA}^{max} = t_{detfix} \frac{\overline{m}_L^L E^L}{\overline{m}_B^M E^M} \tag{46}$$

The maximal zone length for the resolved constituent A, on a time base
is given by $t_{nA}^{max} - t_{detfix}$. The maximal sample load, n_A^{max}, is given
by the intercept of the resolution lines A/B, A/AB and AB/B. For
the given pair of constituents, formate and glycolate, the maximal
sample load was 113 nmole, which was close to the theoretical value
(Table IV). From the maximal sample load the load capacity, C_{load}, and
the dimensionless separation number, S, can be calculated directly.
Optimal column dimensions can be obtained from the load capacity,
whereas the dimensionless separation number gives the relationship
between amount sampled and electrical driving current or time for
resolution.

For anionic straight pairs of constituents, where $m_B < m_A$ and $pK_A < pK_B$
the rationale for optimization is straightforward: low pH of the
leading electrolyte and the sample. With anionic reversed pairs,
where $m_B < m_A$ and $pK_A < pK_B$, this rationale is more complicated. It
has been shown that for such pairs a pH will exist, pH^{MO}, at which
no separation occurs. Of course, this pH will cause an infinite time
for resolution, zero separation number and zero load capacity. Moreover, at this critical pH the order on which the constituents migrate
will be reversed. Experimental results concerning the load capacity

for a reversed pair are given in Fig. 3 as a function of the pH of
the leading electrolyte and the sample. The experimental curves confirm the theoretically predicted behaviour. Using the appropriate
data (Table V) and relevant mathematical formulations, it follows that

TABLE V

Data for a reversed pair of sample constituents

Constituent	Mobility $(cm^2/V \cdot sec)$	pK	Concentration (M)	pH^{sample}
Acetate	$-41 \cdot 10^{-5}$	4.75	-0.005	4.75
Naphthalene-2-sulphonate	$-30 \cdot 10^{-5}$	0	-0.005	4.75

No separation at $pH^{MO} = 5.19$

No separation at $pH^{L} = 4.98$ ($\rho = -2, m_c = 30 \cdot 10^{-5} cm^2/V \cdot sec$)

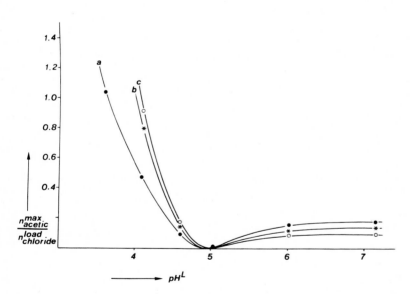

Fig. 3. Influence of the pH of the leading electrolyte on the load
capacity for a reversed pair of constituents. Operational
systems: Table II. Constituent data: Table V. Sample: acetate-
naphthalene-5-sulphonate. (a) $pH^{sample} = 7.0$; (b) $pH^{sample} = 4.75$; (c) $pH^{sample} = 3.00$.

the criterion for separation will not be satisfied at a mixed zone
pH of 5.19. Obviously, this pH can be generated by numerous combina-
tions of leading electrolytes and sample compositions. Working at the
maximal buffering capacity of the common counter constituent, i.e.
$\rho = -2$ or $pH^L = pK_c$, and introducing an acceptable ionic mobility for
the counter constituent, $m_c = 30.10^{-5}$ cm^2/V.sec, the critical pH of
the leading electrolyte, at which no separation occurs, is 4.98.
This was confirmed experimentally by the separation at $pH^L = 5.04$,
at which hardly any load capacity was present.

At a pH^L higher than the critical value, sample constituents migrate
in order of ionic mobilities, and separations can be performed with
only moderate efficiency. At low pH^L, however, constituents migrate
in the order of their pK values and a much greater efficiency can be
obtained, resulting in a high load capacity. For example, the resolution of a 1.5-nmole sample (an absolute amount that can be detected
without difficulty) would take about 18 sec, $S = 0.26$, at $pH^L = 4.10$
and $pH^S = 3.00$, whereas the same sample can be resolved in 105 sec,
$S = 0.045$, at $pH^L = 7.10$.

The required length of the separation compartment in the former
instance is 5.8 times shorter than in the latter. Obviously, for
specific samples rigid optimization procedures can be followed,
resulting in very short analysis times, small dimensions of the
separation compartment and efficient use of the power applied. It
must be emphasized, however, that the success of optimization procedures depends largely on the physicochemical characteristics of the
species to be separated and the performance of the equipment. When
there are only small differences in ionic mobilities and dissociation
constants, optimization procedures are elaborate and result in only a
small increase in efficiency.

From the theoretical formulations it follows that the measurement
of steady-state effective mobilities can only be used as an indication
of experimental separability. Constituents that have equal steady-
state effective mobilities can sometimes be separated efficiently.
Moreover, it has been shown that enforced isotachophoretic configurations (ref. 16) in which a more mobile constituent migrates behind
a less mobile constituent, are stable with respect to time. An example
is given in Fig. 4.

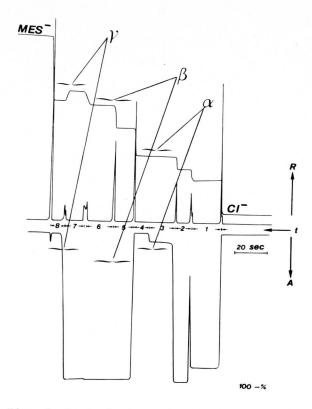

Fig. 4. Isotachophoretic steady-state configurations. Operational system: Table II, system N° 2, pH^L = 3.95. R = resistance; A = UV absorption at 254 nm; t = time , 1 = Adenosine-5-triphosphoric acid (A5TP); 2 = sulphanilic acid; 3 = DL-mandelic acid; 4 = DL-lactic acid; 5 = guanosine 5-monophosphoric acid (G5MP); 6 = adenosine 3-monophosphoric acid (A3MP); 7 = adenosine 5-monophosphoric acid (A5MP); 8 = acetic acid. α = A pair of constituents for which conductimetric detection indicates no resolution whereas UV detection does; β = A pair of constituents for which UV detection indicates no resolution whereas conductimetric detection does; γ = A pair of constituents in an enforced isotachophoretic configuration.

Several sample constituents confirm the general principle that constituents in isotachophoresis migrate at equal velocities in order of decreasing effective mobilities. The constituents lactic acid and mandelic acid (constituents 3 and 4), however, show virtually no difference in effective mobilities; as for their isotachophoretic migration the same electrical gradient seems to be necessary (Fig. 4 and Table VI)? From the linear trace it appears that this pair is not resolved during the separation process. The UV trave however,

TABLE VI

Comparison of zone characteristics

Trans = computerized transient-state model; X3 = computerized steady-state model.

Parameter	Chloride zone, experimental	Mandelate zone			Lactate zone		
		Trans	X3	Experimental	Trans	X3	Experimental
pk	− 2	3.37			3.86		
m(cm^2/V . sec)	−77 . 10^{-5}	−28 . 10^{-5}			−33 . 10^{-5}		
pH	3.95	4.21	4.22	4.25	4.27	4.28	4.29
E^L/E^L	1.00	0.319	0.320	0.322	0.318	3.319	0.322
c_X^X (mM)	−10.00	−6.47	−6.34		−7.16	−7.06	
$\dfrac{\bar{m}_{mandelate\ zone}}{\bar{m}_{lactate}}$	.	0.956	0.960	0.974			
$\dfrac{\bar{m}_{mandelate\ zone}}{\bar{m}_{mandelate}}$					1.027	1.017	1.010
$\dfrac{\bar{m}_{lactate\ zone}}{\bar{m}_{mandelate}}$	1.027						
$\dfrac{\bar{m}_{lactate\ zone}}{\bar{m}_{lactate}}$							
$\dfrac{\bar{m}_{mixed\ zone}}{\bar{m}_{mandelate}}$							
$\dfrac{\bar{m}_{mixed\ zone}}{\bar{m}_{lactate}}$							

indicates clearly that mandelic acid (3) has been resolved from lactic acid (4) and that the former migrates in front of the latter. The transient-state model reveals that the pH of the mixed zone, from which the pure zones are formed, is just below the critical pH of 4.32 at which no separation occurs. As this is a reversed pair, mandelic acid will be resolved in front of lactic acid.

From the data in Table VI it follows that the experimental and theoretical zone characteristics are in good agreement. The minor difference between the transient-state and the steady-state results has already been mentioned: the deviation from the critical value of unity for the ratio of effective constituent mobilities. Mandelic acid ions, in the resolved lactate zone, would lead to a 1 % deviation. From the UV trace it follows that these deviations are large enough to guarantee a sharp separation boundary. The theoretical calculations show a greater difference and for the ratio of effective mobilities in the mixed zone a 2.6 % deviation from unity was calculated. The experimental separation confirms that this deviation is sufficient to obtain resolution. It must be emphasized, however, that the small deviation results in a low separation efficiency, and column overloading can easily occur. Fig. 4 nevertheless indicates clearly that isotachopherograms in which only one universal detection system is used must be interpreted with great care. The same applies, of course, when only UV detection is used. From the UV trace in Fig. 4 it would be concluded that the nucleotides G5MP and A5MP have not been resolved. The conductance trace, however, clearly confirms the separation of these two constituents. On most occasions small amounts of impurities, with either UV-absorbing or non-UV-absorbing properties, will indicate the separation boundary. Moreover, in this particular instance, a difference is visible when the UV results are being traced in the absorbance mode.

The sample constituents A5MP and acetate (constituents 7 and 8) are migrating in an enforced isotachophoretic configuration. The effective mobility of the acetate constituent in its proper zone is higher that that of the nucleotide A5MP in its proper zone, as indicated in Fig. 4 by the lower conductance of zone 7 in comparison with zone 8. For the relative effective mobilities it follows that $m^{-Acetate}_{Acetate}/m^{-Chloride}_{Chloride} = 0.212$ and $m^{-A5MP}_{A5MP}/m^{-Chloride}_{Chloride} = 0.198$.

The 7 % deviation from unity of the mobility ratio ($m^{-A5MP}_{A5MP}/m^{-Acetate}_{Acetate} = 0.93$) allows a satisfactory sharpness of the separation boundary between the two constituents. The reason for the stability can be found in the difference in the pH value of the two resolved zones. Using the appropriate relationship it follows that the pH of the acetate zone is 4.57. A nucleotide ion, lost by convection or diffusion from its proper zone (7) into the acetate zone, will migrate with a higher effective mobility than that of the acetate constituent. In the nucleotide zone the pH is 4.32, so any acetate ion in the nucleotide zone will migrate with a considerably lower velocity than the nucleotide $m^{-A5MP}_{A5MP}/m^{-A5MP}_{Acetate} = 1.37$.

Hence the self-resorting capabilities of the separation boundary allow the enforced isotachophoretic configuration to be stable with respect to time. It should be noted however, that enforced isotacho- phoretic configurations will not be encountered frequently in practice. From both our previous theoretical considerations and the experimental evaluation presented here, it follows that through optimization a considerable increase in separation efficiency and load capacity and a decrease in time for resolution can be obtained. It must be empha- sized that the succes of such a procedure depends largely on the nature of the sample. We restricted our theoretical and experimental studies to two constituent samples but the same optimization rationales hold, to a lesser extent, for multiconstituent samples. For very complex mixtures, in which multicomponent information must be obtained, optimization can sometimes be elaborated and difficult. Analyses in more than one operational system are inevitable. Moreover, in one- component analyses of multiconstituent samples a considerable amount of effort is put into the separation of constituents of little interest. Column-switching techniques (ref. 28) may prove useful here.

4. QUALITATIVE AND QUANTITATIVE ASPECTS
4.1 Introduction

As shown in fig. 5, the signals derived from a conductivity detector (potential gradient, thermometric) possess general characteristics. From these signals qualitative and quantitative information can be deduced. Various ways can be given to list the qualitative information such that laboratories working with other equipment can make use of it. For a correct qualitative evaluation it is always important to known whether the temperature difference between leading electrolyte and terminating electrolyte is large or small; for example,the temperature difference between the chloride zone and the morpholine-ethane sulpho- nate zone of Table VII. The final temperature not only influences the effective mobility (2 % per °C), but also the pK_a values, especially of cationic species. For quantitative evaluation the temperature is of less influence, as long as calibration curves are used. If ionic species are present that migrate faster than the leading ion this can be seen in the linear signal of the conductivity detector, which indicates a conductivity lower than that of the leading ion. The qualitative information of all other ions, migrating between the leading zone and the terminating zone, is not lost. If desired a leading ion can be chosen with a higher effective mobility, in order

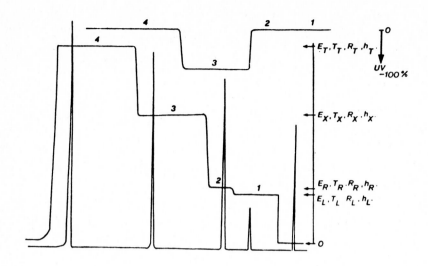

Fig. 5. Isotachophoretic separation of perchlorate and p-aminobenzoate, carried out in the system listed in Table II (5). For explanation see text. 1 = chloride; 2 = perchlorate; 3 = p-amino-benzoate; 4 = MES.

TABLE VII

Survey of the detectors currently used in analytical isotachophoresis

Type	Performance	L_{min} mm	Q_{min}	t_{av} min
Thermal	Thermocouple 15 μm (Cuconstantan) Thermistor (Philips micro beat)	5	$1/2 \ 10^{-9a}$	30
Conductivity	Microsensing electrodes (1-20 kHz)	0.05	10^{-12b}	10
Potential gradient	Microsensing electrodes in combination with the AC-DC converter (ref. 25)	0.05	10^{-12b}	10
UV	Microwell wavelengths: 256,280 nm	0.05	10^{-12b}	10

L_{min} = minimum detectable zone length;

Q_{min} = minimum detectable amount of ionic component in g. equivalents;

t_{av} = average time for analysis.

[a] In = g. equivalents (0.4 mm I.D. narrow-bore-tube).

[b] In = g. equivalents (0.2 mm I.D. narrow-bore tube).

to have all sample constituents between the leading and terminating
zones. If an ion is present in the sample with an effective mobility
equal to that of the leading ion the conductivity of the leading
electrolyte is not changed. Quantitative information can still be
obtained by measuring the retardation of the appearance of the first
separation boundary. Of course in this case the conductivity and zone
lengths of all other zones are not influenced by this ionic species.

4.2 Qualitative aspects

Fig. 5 is an isotachopherogram that was obtained with a potentio-
metric detector. A comparable result can be obtained from a thermometric detector, or a conductimetric detector. For this reason not
only the potential gradient (E) is indicated, but also the values
for the conductimetric detector (R) and the thermometric detector
(T). In this figure is shown the linear trace of the potential gradient
detector, the differential of this and the linear trace of the UV
absorption detector. Qualitative information is derived from the
linear trace of the potential gradient detector. The quantitative
information is obtained by measuring the step lengths (distances
between peaks of the differential trace). The UV absorption detector
gives additional qualitative (specific) information.

Four different methods are given for listing the qualitative data
in operational systems:

(i) The step height h_x;

(ii) The reduced step height $h_x - h_L$;

(iii) The step height unit value or S.U. value

$$100 \frac{h_x}{h_L};$$

(iv) The reference unit value or R.U. value

$$100 \frac{h_x - h_L}{h_R - h_L}$$

From these possibilities h_x gives obscure information, especially
if qualitative information, obtained from the various detectors, is
compared. The conductimeter indicates R_L, which is constant at various
current densities, assuming the temperature differences are small

enough. The potential gradient detector and the thermometric detector does not give any signal at I = 0 µA. Both T_L and E_L are determined by the current density applied. T_L has a square relation and E_L has a linear relation with the current density. The step heights and the reduced step heights vary, moreover, if the amplification is changed. The S.U. values, the potential unit values, the temperature unit values and the conductivity unit values are applicable as they give a ratio in $\mu_{effective}$. Therefore, we prefer to work with these S.U. values in practice. Moreover, a correction is made for the amplification of the electrical circuitry applied, although this holds also for the R.U. value. The R.U. value is less dependent on the current density than the S.U. value, because two reference species are considered. Therefore the accuracy of an R.U. value can be greater than that of an S.U. value.

For listing qualitative data it is always important that the conditions of the operational electrolyte systems applied are well defined, the current density is fixed, the output signals of the detectors are linearized and, for high accuracy, the diameter of the narrow-bore tube is well chosen. We found the optimum for this diameter to be 0.2 mm I.D.

4.3. Quantitative aspects

As Fig. 6 shows quantitative information from isotachophoretic analyses can be obtained by measuring the zone lengths. There is a linear relationship between the zone length of an ionic species and the amount of that ionic species introduced as a sample, assuming the electric current is stabilized. The calibration curves of these ionic species are given in Fig. 7.

If mobility and pK_a values are known, a calibration constant can be determined (ref. 16). This is a constant in each operational system, if the operational conditions are fixed, for all ionic species. The use of an internal standard for quantitative evaluation is described in ref. 16.

At present we prefer to work with calibration curves. We found that this is more accurate, because they are determined experimentally and correct for dissociation, complexation, temperature, solvation and activity, for example.

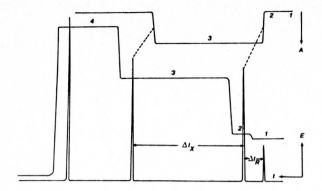

Fig. 6. Isotachophoretic separation of perchlorate and p-aminobenzoate carried out in the system listed in Table II (5). From this figure (and Fig. 5) data are used for the calibration curve (Fig. 7). Perchlorate was used as internal standard. 1 = chloride; 2 = perchlorate: 3 = p-amino-benzoate; 4 = MES; R = increasing resistance: A = increasing UV absorption and t = increasing time; E = electric field strength.

5. APPARATUS

Analytical isotachophoretic experiments are commonly performed in equipment consisting of a narrow-bore tube of PTFE with inner diameters of 0.2-0.4 mm. Such equipment is discussed extensively in the scientific literature. The equipemnt with coupled columns is described in ref. 28.

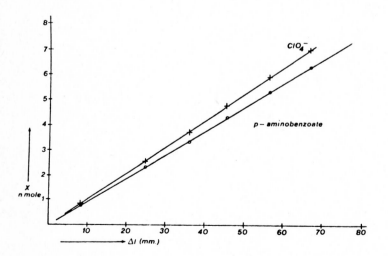

Fig. 7. Calibration curve for quantitative evaluation.

As detectors the conductivity and the UV-absorption detectors are used simultaneously. Commercially available is the Tachophor (LKB Produkter, Bromma, Sweden) equipped with a thermometric and a UV absorption detector. This equipment can be used for micropreparative experiments, using the Tachofrac. The isotachophoretic analyser (Schimadzu, Oakland Ridge Industrial Center, 9147-H Red Road, Columbia, Md. 21045,USA) is equipped with a potential gradient and UV absorption detector.

The experiments described in this paper were performed in home-made equipment (Eindhoven University of Technology, Department of Instrumental Analysis). Table VII gives some general characteristics of detectors, currently used in analytical isotachophoresis. Nearly all values, especially the minimal detectable amounts, refer to the equipment, using a narrow-bore tube of PTFE having an inside diameter of 0.2 mm and an outside diameter of 0.35-0.40 mm.

6. FIELDS OF APPLICATION

It is difficult to describe all possible fields of application of analytical isotachophoresis. To predict the feasibility for isotachophoresis as an analytical method, the ratio of molecular weight to effective charge can be used. Generally this ratio should not exceed 300, although examples can be given in which it can exceed 3000. Needless to say the compound must have sufficient solubility in the solvent chosen. A brief survey of possible fields will be given: amino acids, carbohydrates, complexes, diseases, enzyme assays, inorganic anions, metal ions, nucleotides, organic ions, peptides and proteins.

More about fields of applications can be found in this volume, which essentially is devoted to the applications of isotachophoresis.

LIST OF SYMBOLS

a	degree of dissociation
A	constituent to be separated
B	constituent to be separated
$\bar{c}$	constituent concentration (mole/cm^3)
c	subspecies concentration (equiv./cm^3)
C	counter constituent
D	diffusion coefficient (cm^2/sec)
E	electrical field strength (V/cm)
F	Faraday constant (C/equiv.)
i,j	constituent, subspecies
J	electrical current density (A/cm^2)
κ	electrical specific conductance (Ω^{-1}.cm^{-1})
k	reduced mobility
K	dissociation constant
L	leading constituent
l	zone length (cm)
m	mobility (cm^2/V.sec)
$\bar{m}$	effective mobility (cm^2/V.sec)
n	subspecies, A, B, C, L, T
n	amount of constituent (mole)
O	area (cm^2)
pH	pH
pK	negative logarithmic transformation of K
ϕ	sampling ratio
χ	molar sampling ratio
r	relative mobility
R	resolution
ρ	relative leading concentration
S	separation number
T	terminating constituent
t	time coordinate (sec)
t_{res}	time of resolution (sec)
t_{det}	time of detection (sec)
t_{detfix}	running time (sec)
v	linear velocity (cm/sec)
X	plate coordinate (cm)
X_{res}	length of resolution (cm)
X_{det}	length of detection (cm)
X_{detfix}	running length (cm)

Subscripts

i, j, n A, B, L, T, C constituent, species indicator

Superscripts

K A, B, L, T, M (mixed) zone indicators

**   terminating compartment

*   sampling compartment

                  separation compartment

REFERENCES

1 O. Lodge, Brit. Ass. Advan. Sci. Rep., 56(1886)389.
2 W.C.D. Whetham, Phil. Trans. Roy. Soc., London, Ser. A., 184(1893) 337.
3 W.C.D. Whetham, Phil. Trans. Roy. Soc., London, Ser. A., 186(1895) 507.
4 F. Kohlrausch, Ann. Phys. (Leipzig), 62(1897)209.
5 J. Kendall and E.D. Crittenden, Proc. Nat. Acad. Sci. U.S., 9 (1923)75.
6 D.A. MacInnes and L.G. Longsworth, Chem. Rev., 11(1932)171.
7 A.J.P. Martin, unpublished results, 1942.
8 B.P. Konstantinov and O.V. Oshurkova, Dokl. Akad. Nauk. SSSR, 148(1963)1110.
9 F.M. Everaerts, graduation report, University of technology, Eindhoven, 1964.
10 L. Ornstein, Ann. N.Y. Acad. Sci., 121(1964)321.
11 B.J. Davis, Ann. N.Y. Acad. Sci., 121(1964)404.
12 E. Schumacher and T. Studer, Helv. Chim. Acta, 47(1964)957.
13 A. Verstermark, Cons. electrophoresis: An Experimental study, unpublished results, 1966.
14 W. Preetz, Talanta, 13(1966)1649.
15 W. Preetz, and h.L. Pfeifer, Anal. Chim. Acta, 38(1967)143.
16 F.M. Everaerts, J.L. Beckers and Th.P.E.M. Verheggen, Isotachophoresis - Theory, Instrumentation and Applications (Journal of Chromatography Library, Vol. 6); Elsevier, Amsterdam, Oxford, New York 1976.
17 F.E.P. Mikkers, F.M. Everaerts and J.A.F. Peek, J. Chromatogr., 168(1979)293.
18 F.E.P. Mikkers, F.M. Everaerts and J.A.F. Peek, J. Chromatogr., 168(1979)317.
19 H.A. Hasselbalch, Biochem. Z., 78(1916)112.
20 T.M. Jovin, Biochemistry, 12(1973)871, 879 and 890.
21 J.P.M. Wielders and F.M. Everaerts, in B.J. Radola and D. Graesslin (Editors), Electrofocusing and Isotachophoresis, Walter de Gruyter, Berlin, New York, 1977, p. 527.
22 G. Brouwer and G.A. Postema, J. Electrochem. Soc., Electrochem. Sci., 117(1970)847.
23 R.A. Alberty, J. Amer. Chem. Soc., 72(1950)2361.
24 R. Consden, A.H. Gordon and A.J.P. Martin, Biochem. J., N° 40, (1946)33.
25 Landolt-Börnstein, Zahlenwerte and Funktronen, Spruyer Verlag, Berlin, Göttingen, Heidelberg, 1960, II-7.
26 R.C. Weast (Editor), Handbook of Chemistry and Physics, Chemical Rubber Co., Cleveland, Ohio, 49th ed., 1977.
27 D.D. Perrin, Dissociation Constants of Organic Acids,and Bases in Aqueous Solutions, Butterworths, London, 1965.
28 F.M. Everaerts, Th.P.E.M. Verheggen and F.E.P. Mikkers, J. Chromatogr., 169(1979)21.

A. Adam and C. Schots (Editors), *Biochemical and Biological Applications of Isotachophoresis*   41
© 1980 Elsevier Scientific Publishing Company, Amsterdam — Printed in The Netherlands

A NEW CONSTRUCTION FOR COLUMN COUPLING IN ISOTACHOPHORESIS

Th.P.E.M. Verheggen, F.E.P. Mikkers, D.M.J. Kroonenberg and F.M.
Everaerts
Department of Instrumental Analysis, Eindhoven University of Technology,
Eindhoven (The Netherlands).

SUMMARY

A new construction for selective trapping of isotachophoretic zones,
using a bifurcation, is given. This bifurcation has three channels.
The pre-separation compartment and the separation compartment have
circular dimensions and are in line with each other. In the pre-separation compartment a conductivity detector, the so-called "tell-tale"
detector, is mounted. The third channel is flat. The construction is
carried out in an acrylic block, onto which PTFE capillaries can be
mounted.
An electronic device has been developed for automatic isotachophoretic
analysis, using the coupled column system. This system allows high
sample loads, large concentration differences between the sample constituents and the use of different electrolyte systems.

INTRODUCTION

Samples solutes of interest are often present in solution together
with numerous substances at higher concentrations and thus sample pre-
treatment procedures, such as extraction, column techniques or salting-
out procedures have to be used.

Using the column-coupling system (ref. 1) a large amount of sample
can be introduced, without sample pre-treatment. The equipment described
has a greater potential than simply increasing the maximal load capacity.
Combination with ofther separation techniques, such as liquid chromatography or zone electrophoresis, can be considered.

INSTRUMENTATION

The apparatus (Fig. 1) consists essentially of three sections:

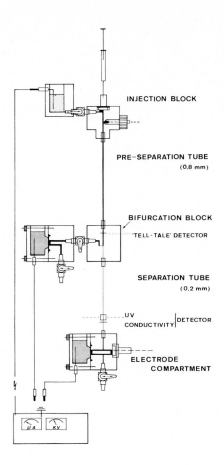

Fig. 1. Equipment with coupled columns suitable for isotachophoretic
analysis. The pre-separation and final separation compartments
are in line with each other. Diffusion effects and disturbance
of the boundary profiles, mixing, are minimized.

(i) The pre-separation compartment; (ii) the bifurcation block with
the"tell-tale" detector; (iii) the final separation compartment. Besides
the bifurcation block, all other compartments and the way of mounting
the detectors and the capillaries are extensively discussed in ref. 2.
The bifurcation block consists of three different channels: (i) A wide
bore of, for example 0.8 mm, in which the "tell-tale" detector is mounted. This bore is connected with a wide-bore PTFE pre-separation tube.

(ii) A narrow bore of, for example 0.2 mm, in line with the wide bore.
Onto this bore the narrow-bore PTFE separation tube is mounted. (iii) A
flat channel (1 mm wide and 0.05 mm high) perpendicular to the bores
of 0.8 mm and 0.2 mm. This channel forms the connection with the counter-electrode compartment, applied during the pre-separation.

The dimensions of the flat channel are chosen in such a way that
diffusion effects are minimized and optimal trapping is allowed.
Since the separation compartments are in line with each other, there
is only minimal mixing of separated zones during trapping. It should
be emphasized that the distance between the pre-separation and separation compartments in the bifurcation block is only 0.05 mm. Because
the heat transfer in this channel is excellent, high pre-separation
currents are permitted.
The "tell-tale" detector identifies the pre-separation. Because the
zones, under carefully chosen operational conditions, all migrate with
equal velocity, the time needed for a zone boundary to cover the distance between the "tell-tale" detector and the bifurcation is constant.
The delay time, t_1, is readily determined using dyes or by measuring
dV/dt of the current-stabilized power supply in a leading electrolyte/
terminating electrolyte experiment.
Once the zone(s) have been idenfitied by the "tell-tale" detector,
the moment at which they reach the branching of the pre-separation
column into the analytical column is well known. Hence the zone(s)
of interest can be easily selected, even if they do not migrate con-
secutively. These zone(s) are further analysed and detected in the
analytical column by both the UV and the conductivity detector.
The maximal load capacity can be increased by an order of magnitude
without increasing the total analysis time. The automation of the
column-coupling system is shown in Figs. 2 and 3. Before the analysis
is started the driving currents for the pre-separation (I_1) and final
separation (I_2), the delay-time t_1 (via TS1), the separation time t_2
(via TS2) and the recording time t_3 (via TS3), and the conductivity
level of the selected zone for the level comparator must be chosen.

As soon as the analysis is started the counter A is activated in
combination with the display. The various times can be measured with
this counter. As soon as the signal derived from the "tell-tale"
detector matches the pre-set value in the level comparator, counter
B is activated. In the memory of this counter the delay time t_1 is

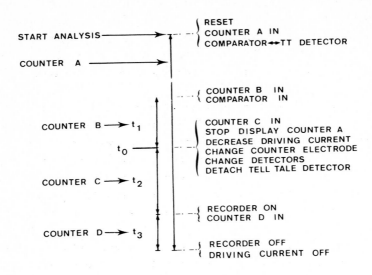

Fig. 2. Important procedures for automatic handling of the equipment with coupled columns. t_0 = pre-separation time; t_1 = delay-time; t_2 = final separation time; t_3 = recording time.

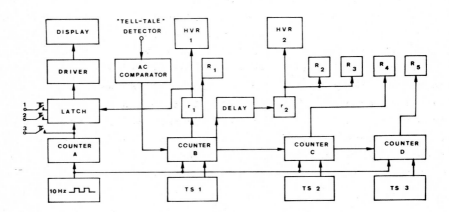

Fig. 3. Schema of the electronic device for automatic analyses with the coupled-column system. For explanation see text. The high-voltage relays were purchased from Kilovac Corporation (P.O. Box 4422, Santa Barbara, Calif. 93103, U.S.A.): HVRI = H26 break-make, 15 kV); HVR2 = H24 (make, 30 kV).

stored via the thumbwheel switch TS1.

After t_1 seconds the following procedures are started simultaneously: (i) Counter C is activated; (ii) The latch is activated, which stops the display of counter A; (iii) The driving current is lowered from its value I_1 to I_2 via the relays R1; (iv) The analytical column is connected to the current-stabilized power supply via the relay HVRI (make-contact 15 kV).

After 100 msec the first counter-electrode (used in the pre-separation mode) is disconnected via the high-voltage relay HVR2 (break-contact 30 kV). Simultaneously, the "tell-tale" detector is mechanically (motor driven) disconnected via relay R2 from the coupling transformer (ref. 2), which has a galvanic insulation up to ca 6 kV.[+] Moreover, the electronics of the conductivity detector are switched from the "tell-tale" detector towards the conductivity detector in the analytical column, via relay R3.

The display of counter A can be activated again manually by de-activating the latch. The pre-separation time, t_0, at which the display stops, can provide valuable information about the amount of ionic material that passed the bifurcation in the pre-separation mode.

The final separation time in the analytical column is stored in the memory of counter C, via the thumbwheel-switch TS3. After this time has passed, the paper transport of the recorder is started via relay R4. Simultaneously counter D is activated. In the memory of this counter the recording time, t_3, is stored via the thumbwheel-switch TS3. After t_3 seconds the recorder is stopped and the driving current (I_2) is switched off. The equipment is now ready for resetting.

EXPERIMENTAL

The equipment with the coupled columns was tested and compared with conventional equipment (refs. 1 and 3-7). Straight calibration curves, injected amount versus zone length, were obtained and no loss of material due to the construction of the bifurcation block could be measured.

+ Note: At the end of the analysis the "tell-tale" detector may reach a potential of approx. 15 kV towards ground level.

The equipment proved to be especially useful when ionic species
in the urine and serum level need to be analysed, as can be found in
refs. 3-7.

Reproducibilities, even in the day-to-day variation, to within
less than 2 % were obtained.

REFERENCES

1 F.M. Everaerts, Th.P.E.M. Verheggen and F.E.P. Mikkers,J.Chromatogr.,
  169(1979)21.
2 F.M. Everaerts, J.L. Beckers and Th.P.E.M. Verheggen, Isotachophoresis, Elsevier, Amsterdam, Oxford, New York, 1976.
3 F. Oerlemans, Th.P.E.M. Verheggen, F.E.P. Mikkers, F.M. Everaerts
  and C. de Bruyn, Advan. Exp. Med. Biol., in press.
4 Th.P.E.M. Verheggen, F.E.P. Mikkers, F.M. Everaerts, Protides Biol.
  Fluids, Proc. Colloq., 27(1979) in press.
5 Th.P.E.M. Verheggen , F. Oerlemans, F.E.P. Mikkers, F.M. Everaerts
  and C. de Bruyn, J. Chromatogr., in preparation.
6 F. Oerlemans, Th.P.E.M. Verheggen, F.E.P. Mikkers, F.M. Everaerts
  and C. de Bruyn, in Biochemical and Biological application of Isotachophoresis. A. Adam and C. Schots (Editors) Elsevier, Amsterdam,
  Oxford, New York,(1980), p. 63-68.
7 F.E.P. Mikkers, Th.P.E.M. Verheggen, J.A.R.J. Hulsman, C.A.M. Meijers
  and F.M. Everaerts, J. Chromatogr., in preparation.

KININOGENS:

I. PURIFICATION OF HUMAN LOW MOLECULAR WEIGHT (L.M.W.) KININOGEN

A. Adam (1,2), J. Damas (3), C. Schots (4), G. Heynen (2),
P. Franchimont (2)

1 - Centre Hospitalier de Sainte Ode, 6970 BACONFOY - BELGIQUE
2 - Laboratoire de Radioimmunologie (Dir. Prof. P. Franchimont) Tour de
    Pathologie, Batiment B23, 4è étage, Université de Liège, 4000
    LIEGE - BELGIQUE
3 - Laboratoire de Physiologie (Dir. Prof. J. Lecomte) Place Delcour,
    Université de Liège, 4000 LIEGE - BELGIQUE
4 - L.K.B. Produkter, 23, Europalaan, 9820 WESTREM

SUMMARY

    A method for the purification of low-molecular-weight   kininogen
is described. It consists of four steps: chromatography on QAE-Sephadex,
ammonium sulphate precipitation, preparative isotachophoresis, and
affinity chromatography on Blue Sepharose CL-6B.

    The product obtained is biologically active and has an activity
equivalent to 15 μg of bradykinin per mg of protein. The specific
activity is thus about 70 times higher than that of plasma (220 ng
bradykinin per mg of protein). The biochemical homogeneity of the
kininogen prepared in this way has been tested by analytical isotachophoresis, molecular sieving, polyacrylamide gel electrophoresis
with and without SDS, and immunoelectrophoresis.

INTRODUCTION

    The kinins play a major role in the genesis and termination of
various physiological and pathophysiological processes, such as increase in vascular permeability, smooth muscle contraction, production of
the inflammatory reaction, blood coagulation and fibrinolysis (refs.
1-3). These biologically active peptides arise from circulating plasma
precursors, the kininogens.

Quantitative measurement of kininogens by radioimmunoassay would allow a more precise definition of the pathogenic role of these substances. However, the establishment of such an assay necessitates the preparation of biologically active and chemically pure kininogen. Plasma kininogens are present in two forms of different molecular weights termed low-molecular-weight (LMW) and high-molecular-weight (HMW) kininogens (ref. 4). We have investigated the purification of LMW kininogen using both classic techniques (refs. 5 and 6) and new and original techniques.

At each step in the purification, measurement of bradykinin-like activity has allowed the localisation of biologically active protein fractions. The purity of the material obtained has been checked using polyacrylamide gel electrophoresis with and without SDS, analytical isotachophoresis, chromatography on a molecular sieve, and, finally, immunoelectrophoresis.

MATERIAL AND METHODS

Methods of purification of LMW kininogen

Siliconised glassware was used for the various purification steps. Kininogen was extracted from human plasma in several steps. The initial three steps were derived from the studies of Habal and co-workers (ref. 5) and Komiya and co-workers (ref. 6). The two last steps in the purification are original.

Step 1

A 2-l volume of human plasma was collected by plasmaphoresis (refs.7 and 8) in four donors with normal coagulation indices. To this was added, in an amount corresponding to 1 % of the total volume, a solution containing the following enzyme inhibitors: Trasylol Ⓡ (Bayer) (10^6 UEKI); benzamidine chlorhydrate (8.7 g); EDTA (3.7 g); hexadimetrine bromide (Polybrene Ⓡ) (Aldrich) (1 g); and soybean trypsin inhibitor (SBTI) (Sigma) (1 g/100ml).

The nature and concentration of the different inhibitors were determined by preliminary experiments so as to minimise the loss of biological activity during the various purification steps. The treated plasma was dialysed at room temperature until its conductivity was that of 0.1 M Tris-HCl (pH 8) buffer.

The same enzyme inhibitors in the same concentration were added to various buffer solutions simultaneously, with a bacteriostatic agent, NaN$_3$, at a final concentration of 0.02 % (w/v).

Step 2.

Chromatography on QAE-Sephadex A-50. A column (10cm/m) of QAE-Sephadex A-50.(Pharmacia) was prepared in 0.1 M Tris-HCl (pH 8) buffer. Protein fractionation was carried out at a room temperature using a stepwise molarity gradient: the buffer itself initially, followed by the same Tris-HCl buffer containing NaCl concentrations:

0.075 M - 0.12 M -0.18 M - 0.35 M.

The flow-rate was 180 ml/h. The elution profile was followed by absorption at 280 nm (Uvicord LKB 8300). The return of the optical density to the baseline determined the duration of each step. The bradykinin-like activity of the different protein fractions was measured.

Step 3.

Dialysis and ammonium sulphate precipitation. The biologically active fractions were dialysed against 0.1 M Tris-HCl (pH 8) buffer and were then precipitated separately at 25°C with 45 % saturated ammonium sulphate with continuous stirring for 45 minutes. The precipitates were collected after 30 minutes centrifugation at 8000 g and resuspended in 0.03 M acetate buffer (pH 5.8). After dialysis and centrifugation the supernatants were tested biologically and lyophilised.

Step 4.

Isolation by preparative isotachophoresis (Uniphor LKB 7900). The support was a polyacrylamide gel (70 ml, 3.2 % w/v) prepared in a leading solution of 10^{-2} M of HCl, 2 x 10^{-2} M histidine (pH 6-6.2). The terminating solution was 2 x 10^{-2} M histidine, 10^{-2} M diethylbarbituric acid (final pH 7-7.2). The sample that was run (18 h, 8mA), was an aliquot (50 mg) of the protein fraction obtained from the third step with or without the addition of four different spacers (10 mg, Sigma) as follows: cacodylic acid, (2-acetamido) - 2 aminoethane-sulphonic acid (ACES), morpholinopropane sulphonic acid (MOPS) and N-2-hydroxy-ethylpiperazine N'-2-sulphonic acid (HEPES).

The various fractions were eluted using a buffer of 2 x 10^{-2} M histidine-HCl (pH 6.7). Optical density at 280 nm was used to follow the elution of the various protein peaks. All fractions were tested by investigating their bradykinin-like activity and the biologically active fractions were studied in various analytical systems to check their purity (CD).

Step 5.
Affinity chromatography on Blue Sepharose CL-6B.
After washing, the Blue Sepharose CL-6B (Pharmacia) (column of 250 mm, diameter 15 mm) was equilibrated with a 0.05 M Tris-HCl buffer (pH 7) containing 0.1 M KCl. The active protein fraction (10 mg) resulting from several isotachophoretic runs was initially dialysed against the same buffer. Chromatography was carried out initially with the equilibration buffer and later with the same buffer containing 1.5 M HCl. A flow-rate of 6ml/h was maintained during this step.

Protein assay
Protein concentrations were measured with Folin-Ciocalteu reagent using the technique described by Lowry (ref. 9).

Measurement of bradykinin-like biological activity
The kininogens were measured by the bioassay of the bradykinin which was released following exposure to trypsin (ref. 10). All samples to be tested (1 ml) were incubated for 37° at 30 min with 0.5 ml of trypsin (250 µg/ml). They were subsequently added to a segment of guinea-pig terminal ileum and to rat duodenum continuously superfused with oxygenated Tyrode's solution (95 % O_2-5 % CO_2) maintained at 37° with added atropine and promethazine (5 x 10^{-7} M), the organs having previously been desensitized to the muscle-stimulating action of trypsin.

The measurement was carried out by comparing the activities of the samples with a curve generated with pure bradykinin (10-200 ng/ml, bradykinin triacetate, Sigma). The results were expressed in ng equivalents of bradykinin per ml or per mg of protein.

Determination of purity and molecular weight of the kininogen
The degree of purity of the kininogen was assessed by five techniques:

(a) Analytical isotachophoresis (LKB Tachophor 2127). The solutions to be tested were adjusted to a concentration of 500 µg/ml, the solutions being used either alone or after the addition (1 µmole/ml) of the four different spacers already mentioned.

Ten microliters of sample were injected into a capillary (60 cm long, 0.5 mm internal diameter) previously filled with a leading solution of 10^{-2} M HCl, 2×10^{-2} M histidine, 0.4 % (w/v) hydroxymethylpropylcellulose (pH 6-6.2). The terminating solution was of the composition described above.

After an initial phase of equilibration (V = 10 kV), migration was carried out at a constant strength of 115 µA, the voltage increasing progressively from 10 to 30 kV. The different proteins were detected by microcalorimetry and by optical density at 280 nm.

(b) Polyacrylamide gel electrophoresis using the technique of Davis (ref. 11). The protein samples with the addition of Bromothymol Blue were subjected to electrophoresis in duplicate in a gel of 2 ml pH 8. The first of these gels was stained with Coomassie Blue and the R_F of each band was calculated. The second gel was sliced into pieces 2 mm thick, each being separately eluted with the electrophoresis buffer bor 24 h at 4°C. The biological activity of each protein fraction was tested.

(c) Electrophoresis in SDS-polyacrylamide gel. The technique described by Laemli (ref. 12) was used. The acrylamide gels (gradients of 5-25 %) were poured in slabs. Electrophoresis was carried out for 45 min at 30 V and then for 5 h at 210-250 V. After staining with Coomassie Blue the distances migrated were compared with molecular weight markers (Combiteck®, Boehringer), allowing molecular weights to be calculated.

(d) Chromatography on the molecular sieve, Sephadex G-200. Chromatography was carried out on a column (1 m long diameter 2 cm) of Sephadex G-200 in 0.1 M Tris-HCl buffer (pH 8). The column had previously been calibrated with molecular weight markers (Combiteck®, Boehringer) and calculation of the K_D allowed an estimate of the molecular weight of the kininogen.

(e) Immunoelectrophoresis. This was carried out using the technique of Grabar and co-workers (ref. 13).

The antisera used were: (1) An antiserum against total human serum protein (Wellcome N° IP 14), and (2) Two antikininogen antisera raised in rabbits using the immunisation scheme of Franchimont (ref. 14)

The antigen injected was the biologically active fraction obtained by isotachophoresis.

RESULTS

The use of enzyme inhibitors resulted in the conservation of biological activity of 1300 ng bradykinin per ml after dialysis compared with 1600 ng/ml before (Table 1).

TABLE 1

Bradykinin activity in the different biologically active fractions at each step of the purification of LMW kininogen

	Volume (ml)	Protein (mg/ml)	Bradykinin µg/ml	Bradykinin µg/mg protein	Bradykinin total (µg)
(A) Plasma	2,000	70	1.6	0.023	3,200
(B) Dialysate	2,300	66	1.3	0.020	2,990
(C) QAE-Sephadex A-50 - Precipitation with 45 % ammonium sulphate					
(1) Protein eluted with 0.12 M sodium chloride (LMW kininogen)					
B1	200	2.18	2.6	1.2	520
B2	200	1.68	2.5	1.1	500
B3	100	2.66	1.9	0.7	190
B4	50	2.10	1.3	0.6	65
(2) Protein eluted with 0.36 M NaCl (HMW kininogen)					
B5	50	2.40	1.80	0.75	375
B6	150	4.60	2.60	0.56	390
(D) Preparative isotachophoresis (50 mg of B1)					
C1	6	0.710	5.60	8	33.6
(E) Blue Sepharose CL-6B (10 mg of C1)					
D1	2	2.65	40	15	80

Figure 1 shows the various purification steps applied to the biologically active fractions.

Chromatography on QAE-Sephadex A-50 and ammonium sulphate precipitation

Fig. 2 illustrates the stepwise elution of plasma dialysate by increasing concentrations of sodium chloride in the 0.1 M Tris-HCl (pH 8) buffer. A first fraction was eluted by this buffer alone; four other protein fractions were eluted successively by the progressively increasing NaCl concentration.

Only the fractions eluted with 0.12 M and 0.35 M NaCl contained bradykinin-like biological activity. These two fractions corresponded to LMW and HMW kininogen, respectively (ref. 1).

On the basis of the four regions of 280 nm absorption, the fraction eluted with 0.12 M NaCl (corresponding to LMW kininogen) was subdivided into four subfractions, each biologically active.

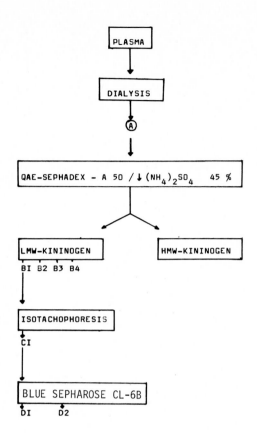

Fig. 1. General scheme of the purification of LMW-kininogen

Likewise, the fraction obtained with 0.35 M NaCl (described as HMW kininogen) also had two subfractions in which we found bradykinin-like activity. Each of these biologically active subfractions was separately precipitated with 45 % saturated ammonium sulphate and resuspended in 0.03 M acetate buffer (pH 5.8).

The results for biological activity and protein concentration are presented in Table 1; the subfractions B1 and B2 eluted with 0.12 M NaCl showing specific activity equal to and greater than the subfractions B3 and B4.

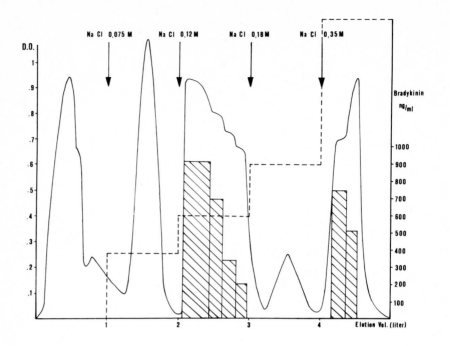

Fig. 2. QAE-Sephadex chromatography. Elution volume is plotted on the
abscissa; biological activity in ng bradykinin per ml (hatched
rectangle) is shown on the right ordinate, optical density
at 280 nm (o———o) on the left ordinate. The dotted lines show
the various concentrations of NaCl in the stepwise gradient.

As shown in Fig. 3, subfraction B1 was found to be very heterogeneous on analytical isotachophoresis. This heterogeneity was confirmed
by SDS polyacrylamide gel electrophoresis. Fraction B1 was therefore
purified further.

Preparative isotachophoresis of LMW kininogen

When subfraction B1 was subjected to preparative isotachophoresis
using the conditions described, we systematically and reproducibly
found bradykinin-like activity in protein fractions eluted between
the two discrete spacers ACES and MOPS (Fig. 4). The specific activity
of this fraction was 8 μg bradykinin per 710 mg protein (Table 1).

The degree of purity of this fraction was examined by analytical
isotachophoresis (Fig. 5), polyacrylamide gel electrophoresis with

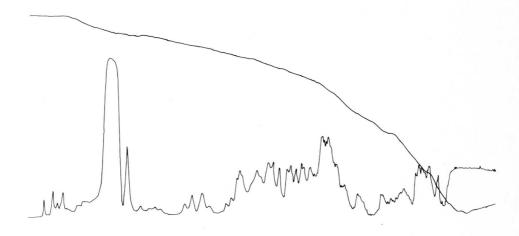

Fig. 3. Analytical isotachophoresis of fraction B1. At the bottom of
the figure the optical density of the protein peaks detected
at 280 nm. At the top, the continuous curve of the thermal
detection.

or without SDS (Fig. 6), and gel filtration, all three techniques
demonstrating the homogeneity of this fraction C1 (Fig. 7). In
contrast, when C1 was subjected to electrophoresis with the corresponding antiserum, two precipitation lines were seen: one in the
albumin zone the other in the globulin zone. The line in the albumin
zone was also seen when the antiserum against human serum protein was
used after electrophoresis of this kininogen preparation. Finally,
when albumin was electrophoresed and then reacted with anti-C1 antiserum, a precipitation line appeared in the corresponding zone of
migration. There are thus two antigens in the biologically active
preparation of LMW kininogen obtained by isotachophoresis: albumin
and a globulin. As human albumin is free of bradykinin-like activity,
the biologically active kininogen must therefore be the antigenic
constituent that migrated in the globulin zone.

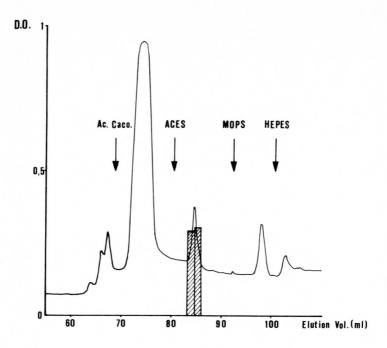

Fig. 4. Preparative isotachophoresis of B1 (50 mg) with the addition
of 10 mg of each discrete spacer. The continuous line shows
optical density at 280 nm. The hatched area shows bradykinin-
like activity.

Affinity chromatography on Blue Sepharose CL-6B

On chromatography of 10 mg of C1 on Blue Sepharose CL-6B the first
protein peak was eluted with a buffer of 0.05 M Tris-HCl (pH 7), 0.10 M
KCl (D1). A second peak (D2) was eluted with the same buffer containing
1.5 M KCl. Biological activity was found only in the first peak. This
protein peak was homogeneous on analytical isotachophoresis, polyacrylamide gel electrophoresis with or without SDS, and on gel filtration
on a molecular sieve. Calculation of the mean K_D gave a molecular
weight between 54,000 and 62,000 for the LMW kininogen, which agreed
well with the results obtained using SDS electrophoresis (Fig. 8).
On immunoelectrophoresis we found only a single precipitation line
in the globulin zone, the line corresponding to albumin having now
disappeared. In contrast, D2 did not release any bradykinin and on
immunoelectrophoresis gave a precipitation line in the albumin zone.

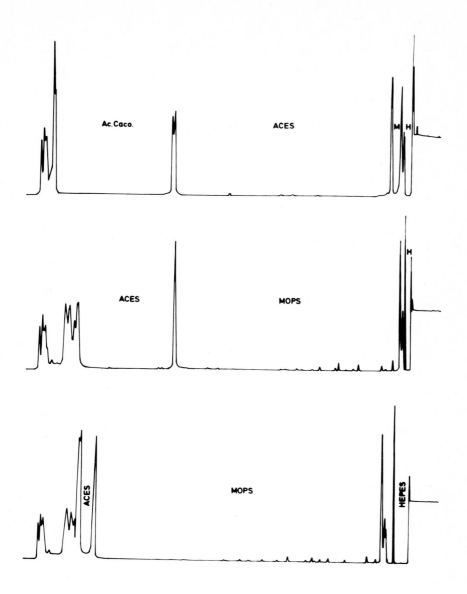

Fig. 5. Analytical isotachophoresis of fraction C1, with the successive addition of different discrete spacers: cacodilic acid, ACES, MOPS, HEPES. ACES and cacodilic acid are shown in the upper section, ACES and MOPS in the middle section, MOPS alone in the lower section of the figure.

58

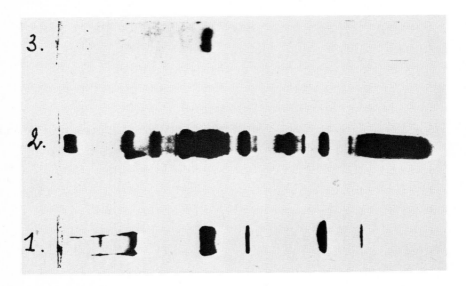

Fig. 6. SDS polyacrylamide gel electrophoresis. 1 = Molecular weight
markers (Combiteck (R)); 2 = Fraction B1; 3 = Fraction C1.

DISCUSSION

The purification scheme that we have established has allowed us
to obtain biologically active LMW kininogen in a pure state. The
first steps using QAE-Sephadex and ammonium sulphate precipitation
gave quantitative results (specific activity) comparable to those of
Habal et al. (ref. 5).

The elution profile was superimposable and biological activity
was comparable. The main advantage of ammonium sulphate precipitation
was to eliminate naturally occuring physiological inhibitors in
plasma and to increase by a factor of approximately 3, the protein
concentration and biological activity whilst retaining specific
activity at a similar level. However,we went further to use isotachophoresis, an analytical and purification technique which has been
developed very recently (ref. 15). We shall not repeat the physico-

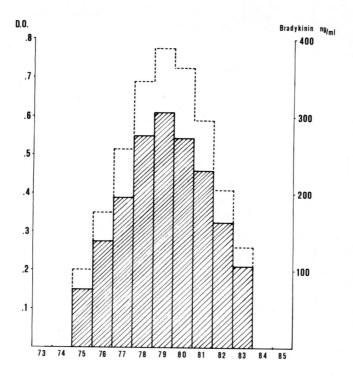

Fig. 7. Chromatography of C1 on molecular sieve Sephadex G-200.
On the abscissa are shown the various fractions eluted; on
the ordinate optical density at 280 nm (continuous line)
and bradykinin activity (hatched bars).

chemical principles on which the technique is based; these have been
described elsewhere (ref. 16).

This rapid technique, in which the analytical steps can readily
be adapted to preparative methods, allowed us to remove the majority
of natural contaminants of kininogen in a single step (ref. 17).
The isotachophoretic step is original from two points of view: the
method has never been used previously and takes advantage of a
physicochemical property of LMW kininogen which has not previously
been exploited, its clear-cut mobility in a defined pH zone between

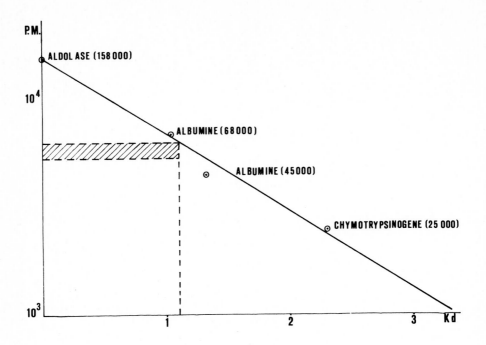

Fig. 8. Molecular weight calculation of LMW kininogen using gel
chromatography. On the ordinate is shown the molecular weight
on the abscissa the K_D of the various markers and of kininogen

the pH of the leading solution (6.2) and of the terminating solution
(7.2). We have thus established a new system of isotachophoresis based
on the pI value of LMW kininogen found in the literature. This pI
was stated to be 3.3 - 5.0 by various authors (refs. 18 and 19).
However, we did not use a system where LMW kininogen would migrate
as a cation at pH less than the pI since kininogen polymerises at
a pH less than pI (ref. 20). We therefore chose an anionic isotachophoretic system.

Here also, the constituents of the system were chosen bearing in
mind the stability of the LMW kininogen molecule which undergoes
self-degradation at an alkaline pH above 8.5 (ref. 21). This is the
reason for the choice of the system described in the Methods section
where the second pK_a of the counter-ion histidine (6.12) fixes the
working zone. In this pH zone, the chloride ion would move more rapidly

(leading), and the diethylbarbituric acid (pK_a = 8.2) would move more slowly, than the kininogen.

The discrete spacers, molecules which have a fixed mobility at a given pH, gave useful separations of the various protein peaks as a result of their intermediate mobility. They did not interfere with the detection of the protein peaks as they fail to absorb at 280 nm and have no biological activity.

ACES and MOPS separated the kininogen from other protein constituents in the isotachophoretic system that we have established. Nevertheless, isotachophoresis in the system used does not free the preparation from albumin which has a pI close to that of kininogen (ref. 17). Some authors (refs. 17 and 22) have used a final purification step of affinity chromatography with anti-HMW kininogen or polyacrylamide gel electrophoresis. They did not, however, specify what contaminants were eliminated (ref. 22). We were able to use Blue Sepharose CL-6B affinity chromatography to eliminate albumin. This problem was not considered by several authors, who confined themselves to purification resulting in increased bradykinin activity. Thus, with a limited number of steps and good yield, we have been able to obtain biochemically pure material which is immunologically homogeneous as shown by the specific activity. Furthermore, this system can be used for the independent purification of HMW kininogen.

ACKNOWLEDGEMENTS

We are grateful to Mrs Duplicy for the practical realisation of this manuscript.

REFERENCES

1 M. Rocha & Silva, (1970), Kinin Hormones with Special Reference to Bradykinin and Related Kinins. Charles C. Thomas Publisher, Illinois.
2 H. Saito, O.D. Ratnoff, R.W. Waldmann, J.P. Abraham, J. Clin. Invest. 55(1975)1082-1089.
3 K.D. Wuepper, D.R. Miller, M.J. Lacombe, J. Clin. Invest. 56(1975) 1663-1672
4 S. Jacobsen, Brit. J. Pharmacol 26(1966) 404-411
5 F.M. Habal, H.Z. Movat, C.E. Burrowes, Biochem. Pharmacol. 23(1974) 2291-2303.
6 M. Komiya, H. Kato, T. Suzuki, J. Biochem. 76(1974)811-822.
7 A. Kliman, M.F. Lesses, Transfusion, 4 N° 6 (1964) 469-472.
8 J.L. Tullis, D.M. Surgenor, R.J. Tinch, M. D'Hont, F.L. Gilchrist, S. Driscoll, W.H. Batchelor, Science 124(1956)792-800.

9 O.H. Lowry, N.J. Rosenbrough, A.L. Farr, R.J. Randall, J. Biol. Chem. 193(1951)265-275.
10 I. Trautsshold, Handbook of Experimental Pharmacology, 25(1974)52.
11 B.J. Davis, Ann. N.Y. Acad. Sci., 121(1964)404-410.
12 V.K. Laemli, Nature 227(1970)680
13 P. Grabar, C.A. Williams, Biochem. Biophys. Acta 17(1955)67.
14 P. Franchimont, Ann. Endocr. 23(1962)556.
15 First International Symposium on "Biological and Biochemical applications of Isotachophoresis". Proceedings in Press (Elsevier 1980). A. Adam and C. Schots Editors.
16 F.M. Everaerts, J.L. Beckers, Th.P.E.M. Verheggen, Isotachophoresis Theory, Instrumentation and applications (Journal of Chromatography Library, vol. 6) Elsevier, Amsterdam, Oxford, London, (1976).
17 U. Hamberg, P. Elg, E. Nissinen, P. Stelwagen, Int. J. Peptides Rés., 7(1975) 261-280.
18 J.C. Londesborough, U. Hamberg, Biochem. J., 145(1975)401.
19 F.M. Habal, H.Z. Movat, Res. Com. Chem. Path. Pharmacol., 4(1972) 477.
20 E. Habermann, Biochem. Z., 377(1963)340.
21 H.J. Freedman, Res. Com. Chem. Pathol. Pharmacol., 15(1976)581.
22 T. Nakayasu, S. Nagasawa, J. Biochem, 85(1797)249-258.

A. Adam and C. Schots (Editors), *Biochemical and Biological Applications of Isotachophoresis*   63
© 1980 Elsevier Scientific Publishing Company, Amsterdam — Printed in The Netherlands

ISOTACHOPHORETIC SEPARATION OF SERUM PURINES AND PYRIMIDINES

F. Oerlemans

Department of Human Genetics, Faculty of Medicine, University of
Nijmegen, Nijmegen (The Netherlands)

Th. Verheggen, F. Mikkers and F. Everaerts
Department of Instrumental Analysis, Eindhoven University of Technology,
Eindhoven (The Netherlands)

and

C. de Bruyn
Department of Human Genetics, Faculty of Medicine, University of
Nijmegen, Nijmegen (The Netherlands)

INTRODUCTION

Purine and pyrimidine metabolism receives attention from a rapidly
growing number of workers in the fields of inborn errors (ref. 1),
hematology (ref. 2), immunology (ref. 3) and oncology (refs. 4 and 5).
The availability of metabolite profiles of body fluids and cell contents
might contribute to a better understanding of mechanisms underlying
metabolic disturbances. This enables a more direct approach for both
diagnostic and experimental purposes. For identification of purines
and pyrimidines thin-layer high-voltage electrophoresis and chromatography can be used (ref. 6). A more rapid technique involves high-
performance liquid chromatography (HPLC) and is widely used at present
(refs. 7 and 8). An alternative to HPLC for screening metabolite profiles might be isotachophoresis (ref. 9). This technique has recently
been introduced for the separation and identification of muscle nucleotides (ref. 10) and urinary purines and pyrimidines (ref. 11). An
advantage of isotachophoresis over HPLC is its flexibility: buffers
can be changed rapidly, no columns need to be equilibrated.

MATERIALS AND METHODS

Uric acid, hypoxanthine, inosine, adenine and guanosine, all of analytical grade, were obtained from Merck (Darmstad, G.F.R.). Orotic acid, hippuric acid, xanthine and allopurinol were purchased from Sigma (St. Louis, Mo., U.S.A.). Hydroxyethylcellulose (HEC) from PolySciences (Warrington, Pa., U.S.A.) was used.

Serum was prepared from venous blood after clotting (2 h at room temperature) and centrifugation for 10 min at 1000 g (4°C). If not analyzed immediately samples were stored at -20°C.

In this study a coupled-column system (ref. 12) was used. The inner diameter of the final separation capillary was 0.2 mm. In Table Ia the electrolyte system for the separation of purine and pyrimidine bases is given. Table Ib gives the electrolyte system for the rapid determination of serum uric acid alone.

TABLE I

Electrolyte system

	System a	System b
Leading ion	Chloride	Chloride
Concentration	0.01 M	0.01 M
Counter ion	0.01 M TRIS[*]/Li[+]	EACA[**]
pH	7.75	5.00
Terminating ion	OH[-]	MES[***]
Additive	HEC[****] 0.25 %	HEC[****] 0.25 %

[*] TRIS = Tris(hydroxymethyl)aminomethane.
[**] EACA = 6-aminohexanoic acid.
[***] MES = 2(N-Morphino)ethane sulfonic acid.
[****] HEC = Hydroxyethylcellulose.

RESULTS AND DISCUSSION

The UV trace in Fig. 1 shows that a standard solution of bases, nucleosides and some other metabolites can be separated with the electrolyte system of Table Ia. It should be noted that the hydroxyl ion was used as the terminating ion. The appropriate terminating solution was made by filtering a saturated $Ba(OH)_2$ solution. Most of the bases, zone N° 5-10 in Fig. 1, differ appreciably in their effective mobilities, since they are separated on the basis of their pK_a values. Orotic, uric and hippuric acids, zones N° 2-4 in Fig. 1, have fairly low pK_a values compared to the pH of the leading electrolyte.

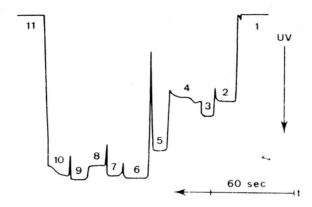

Fig. 1. UV trace from a mixture of bases, nucleosides and some other
metabolites. 1 = Chloride, 2 = Orotic acid, 3 = Uric acid,
4 = Hippuric acid, 5 = Xanthine, 6 = Hypoxanthine, 7 = Inosine,
8 = Allopurinol, 9 = Guanosine, 10 = Adenine, 11 = Hydroxyl.
UV = UV absorption at 254 nm. t : time. Electrolyte system:
system a Table I.

As a result they migrate as ionic species with little difference in
mobility.

    For metabolic studies it is important to profile the metabolites
of interest in serum. At the high pH of the leading electrolyte,
proteins present in a relatively high concentration will obscure the
presence of the nucleosides and bases. Therefore, serum samples have
to be deproteinized before analysis. For the serum studies ultrafiltration through Amicon CF 25.(mol.wt. cut-off 25,000) was used.
A number of solutes were detected in the ultrafiltered serum of a
hypouricemic individual (Fig. 2A). The swamping amount of physiological
chloride, about 300 nmoles, was discarded using the coupled-column
system (ref. 12). As a result the time of analysis did not exceed
15 min. A preliminary identification of the UV-absorbing solutes was
attempted with standard solutions. An example of this is shown in
Fig. 2B, where the analysis of a mixture of the same hypouricemic
ultrafiltered serum, spiked with the standard solution of Fig. 1, is
given. The UV traces of an ultrafiltered normal serum, pooled from
four healthy controls, and an ultrafiltered serum from a patient
with the Lesch-Nyhan syndrome, are given in Figs. 3A and B respectively.

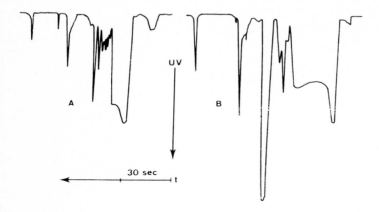

Fig. 2. UV profiles of ultrafiltered sera. A, Ultrafiltered serum of
a hypouricemic individual. B; The same serum "spiked" with
the standard solution used in Fig. 1. UV: UV absorption at
254 nm. t: time. Electrolyte system: system a, Table I.

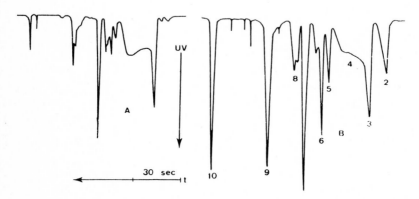

Fig. 3. UV profiles of ultrafiltered sera. A, Ultrafiltered serum
from a healthy control. B, Ultrafiltered serum from a patient
with the Lesch-Nyhan syndrome. UV: UV absorption at 254 nm.
t: time. Electrolyte system: system a, Table I.

In the Lesch-Nyhan syndrome an enzyme of purine metabolism, hypoxanthine guanine phosphoribosil transferase, is deficient. Metabolically
this results in increased levels of uric acid, which was confirmed
by the separation shown in Fig. 3B.

Accurate uric acid determinations can be done with the electrolyte
system b given in Table I.
In a series of experiments isotachophoretic analysis yielded reco-
veries of 99.0-100.5 %. To estimate the amount of uric acid bound
to serum proteins under our experimental conditions the recovery from
ultrafiltered and non-filtered samples was compared. When undialysed
serum was passed through an Amicon CF 25 filter (mol.wt. cut-off:
25,000) 85.1 % of the total serum uric acid was recovered in the
ultrafiltrate, indicating that approximately 15 % was bound to protein
with a mol.wt. exceeding 25,000. The lower amount of uric acid in the
ultrafiltrate compared to non-filtered samples was not due to the
CF 25 filter. The identity of uric acid was confirmed in several ways.
The uric acid zone disappeared after pre-incubation of the sample with
uricase. Injection of an additional, small amount of uric acid increased the length of the uric acid zone. Furthermore, the conductivity
signal (step height) was specific for uric acid.
Comparison of the conventional enzymatic assay of uric acid and isotachophoretic results showed a good correlation.

The usefulness of isotachophoresis in screening for inborn errors
of purine and pyrimidine metabolism by analysing urinary bases and
nucleosides has already been shown (ref. 4). Here and elsewhere (ref.
13) operational systems are presented for the analysis of purine and
pyrimidine nucleosides and bases in serum. For both experimental and
clinical purposes alternative possibilities are opened up; for example,
for pharmacokinetic analysis of drug metabolism, or for studies of the
binding of metabolites such as uric acid to serum proteins.

REFERENCES

1 V.A. McKusick, Mendelian Inheritance in Man, John Hopkins Univ.
  Press, Baltimore, 1975.
2 W.N. Valentine, K. Fink, D.E. Paglia, S.R. Harris and W.S. Adams
  J. Clin. Invest., 54(1974)866.
3 J.E. Seegmiller, H. Bluestein, L. Thompson, R. Willis, S. Matsumoto
  and D. Carson, in F. Hommes (Editor), Models for the Study of Inborn
  Errors of Metabolism, Elsevier/North Holland Biomedical Press,
  Amsterdam, 1979, p. 153.

68

4 E.M. Scholar and P. Calabresi, Cancer Res., 33(1973)94.
5 C. Ip and T. Dao, Cancer Res., 38(1978)723.
6 H.A. Simmonds, Clin. Chim. Acta, 23(1969)353.
7 P.R. Brown, High Pressure Liquid Chromatography - Biochemical and Biomedical Applications, Academic Press, New York, 1973.
8 P.R. Brown, A.M. Krstolevic and R.A. Hartwick, Advan. Exp. Med. Biol., 76A(1977)610.
9 F.M. Everaerts, J.L. Beckers and Th.P.E.M. Verheggen, Isotachophoresis - Theory, Instrumentation and Applications, Elsevier, Amsterdam, Oxford, New York, 1976.
10 D.C. Cower and R.C. Woledge, Sc. Tools, 24(1977)17.
11 H.A. Simmonds, A. Sahota, C.F. Potter and J.S. Cameron, Clin. Sci. Mol. Med., 54(1978)579.
12 F.M. Everaerts, Th.P.E.M. Verheggen and F.E.P. Mikkers, J. Chromatogr., 169(1979)21.
13 F. Oerlemans, Th. Verheggen, F. Mikkers, F. Everaerts and C. de Bruyn, Advan. Exp. Med. Biol.,(1979) in press.

A. Adam and C. Schots (Editors), *Biochemical and Biological Applications of Isotachophoresis*   69
© 1980 Elsevier Scientific Publishing Company, Amsterdam — Printed in The Netherlands

THE DETERMINATION OF VARIOUS LOW-MOLECULAR-WEIGHT CARBOXYLIC ACIDS
IN BIOLOGICAL SAMPLES BY ISOTACHOPHORESIS

J.S. van der Hoeven and H.C.M. Franken

Department of Oral Microbiology, Institute of Preventive and Community
Dentistry, University of Nijmegen, Nijmegen, (The Netherlands)

The determination of low-molecular-weight carboxylic acids including
aliphatic "volatile" fatty acids, hydroxyacids such as lactic and
hydroxybutyric acids, ketoacids such as pyruvic and acetoacetic acids,
dicarboxylic acids, etc., has practical importance in several fields.
In microbiology, the acid end-products of fermentation are analysed
to study metabolic pathways and for the purpose of identifying anaeorbic
bacteria.

Organic acids are assayed in food technology, and in clinical biochemistry the estimation of lactic, acetoacetic and 3-hydroxybutyric
acids serves diagnostic purposes. The methods currently employed to
determine the organic acids are gas chromatography, high-performance
liquid chromatography, or enzymatic assays. Another technique that
can be used is isotachophoresis (refs. 1 and 2). In this paper we
describe the application of isotachophoresis in the determination of
a variety of organic acids in biological samples such as bacterial
cultures, accumulations of bacteria in vivo (for example, dental
plaque), blood, serum and saliva.

EXPERIMENTAL

Apparatus

The apparatus used in this work was the LKB 2127 Tachophor. The
separations were performed in a 20-cm PTFE capillary with 0.4 mm internal diameter. The apparatus was provided with a UV detector and a
conductivity detector (a.c. method) (Fig. 1a,b). The measuring cell
of the conductivity detector consisted of a Perspex cylinder with a
0.4 mm bore along the axis, in which two electrodes of platinum-irridium
(10 %) 10 μm thick were placed at an axial distance of 100 μm. A
conductivity detector is not commercially available yet. The instrument

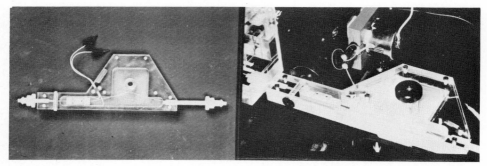

Fig. 1. Conductivity cell and UV cell (a) mounted in a capillary
       block adapted to the LKB 2127 Tachophor, and (b) installed
       in the apparatus with the solenoid close to the conductivity
       cell.

we used we constructed ourselves essentially as described by Everaerts
et al. (ref. 1).

The simultaneous operation of both a UV and a conductivity detector
was found to be useful. The UV detector fails to detect zone boundaries
not marked by UV-absorbing components. On the other hand UV peaks are
occasionally observed, marking zone boundaries not revealed by the
conductivity detector.

Electrolytes

The separations were generally carried out at a concentration of
2.5 mM of the leading ion. This concentration was in accordance with
the concentration of the acidic components to be determined in most
of our samples. If necessary the samples were appropriately diluted.
Alternatively, samples with higher concentrations could be analysed
at higher concentrations of the leading electrolyte. It is noted that
at the low concentration of 2.5 mM, the system can easily be disturbed
because its buffering capacity is low and even small amounts of chemical impurities from various sources are relatively important.

The leading electrolyte consisted of buffered HCl (2.5 mM). The
buffering counter ion was selected according to the desired pH. In
order to obtain maximal buffering capacity, the pK_a of the counter

ionic substance must be close to the pH of the electrolyte. This is
particularly important when the electrolyte concentration and consequently the buffering capacity are low. In most cases the pH of the
leading electrolyte was 3.9 and 4-aminobutyric acid (pK$_a$ 4.03) was
used as counter ion. At pH 4.2 and pH 4.4, ε-aminocaproic acid
(6-aminohexanoic acid, pK$_a$ 4.37) was used as counter ion.
Variation of the pH of the leading electrolyte will result in changes
in the effective mobilities and the migration rate of the ions. This
is shown in Fig. 2a-c, where it can be seen that even small changes
in pH markedly influence the separation.

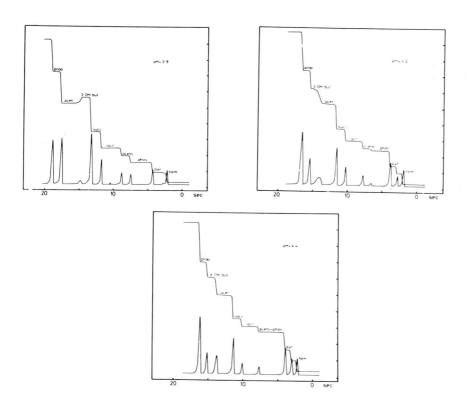

Fig. 2. Isotachopherogram of a mixture containing form, formate; pyr,
pyruvate; phos, phosphate; aceto, acetoacetate; lact, lactate;
succ, succinate; 3-OH-but, 3-hydroxybutyrate; acet, acetate,
and prop, propionate, run at different pH values in 2.5 mM
chloride as leading ion. (a) pH 3.9, (b) pH 4.2 and (c) pH 4.4.
Recorder 200 mm/min.

pH change is applied to identify or confirm the identity of the ionic species. Due to the construction of the apparatus the procedure is very easy and the equipment can be rinsed and filled with new electrolyte in a few minutes.

Polyvinylalcohol (Merck), mol.wt. 72,000 at a concentration of 0.05 %, was added to the leading electrolyte to sharpen zone boundaries by depressing electroendosmosis. The polyvinylalcohol was purified on a column filled with a mixed-bed ion-exchange resin to remove ionic impurities. The terminating electrolyte was 2.5 mM caproic acid that was buffered to pH 5.5 with Tris (2-amino-2-hydroxymethyl-1,3-propanediol). All reagents used were of analytical purity grade. Standard solutions of acid anions were prepared from the respective sodium salts. Deionized-filter sterilized water (Millipore Q) was used to prepare all solutions. The electrolyte solutions were stored at -20°C in portions for daily use.

The electric current was stabilized at 30 µA. The voltage increased from 2.5 kV to approximately 20 kV at the end of the run. The time of analysis was 8-12 min depending on the sample. The chloride content of blood and serum considerably increased the time of analysis.

Analyses of samples

Two strains of oral bacteria, Streptococcus mutans Ny266 and Actinomyces viscosus Ny1, were grown in a complex Trypticase soy broth (BBL) for 24 h at 37°C. The carbohydrate content of this medium is approximately 15 mmoles per liter. The acidic fermentation products were assayed in the cell-free growth medium that was obtained by centrifugation of the bacterial culture for 5 min at 10,000 g.

Dental plaque is the bacterial mass on the tooth surfaces. The metabolism of dental plaque is responsible for the development of caries and periodontal diseases. Since dental plaque is very viscous, it cannot be delivered by syringe. Therefore, the plaque is suspended in histidine (0.1 M/l) which does not interfere with the isotachophoretic analysis of organic acids and which keeps the bacteria viable for subsequent bacteriological counting. Blood, serum or saliva is injected directly into the separation capillary of the Tachophor.

The analysis of biological materials involves a considerable risk of microbial contamination of the apparatus, very often due to the growth of fungi on the walls of the capillary and the electrode compartments. Fungal metabolism resulting in uptake and excretion of a variety of chemicals could seriously interfere with the analysis. Contamination with bacteria from the samples can be avoided by high-speed centrifugation. Extensive rinsing of the apparatus with sterile water is required to prevent microbial growth.

Microbial growth can be removed by ultrasonic treatment in a detergent or with dilute dichromate-sulphuric acid or nitric acid treatment.

RESULTS AND DISCUSSION

The separation of a mixture of organic acids at three different pH values of the leading electrolyte is shown in Fig. 2a-c. The mixture contained formate, pyruvate, acetoacetate, lactate, succinate, 3-hydroxybutyrate, acetate and propionate in slightly different concentrations. Phosphate was added to the mixture because of its prevalence in biological samples. The time of analysis was approximately 12 min. The detector signals were generally recorded at a high speed of 200 mm/min. The steepness of the steps in the conductivity signal indicated good resolution by the detector.

The effect of pH on the isotachophoretic separation is apparent from the changes in the step heights and the zones lengths in the recorder tracings. This is illustrated by the observation that formate and pyruvate are separated at pH 4.2, but not at pH 3.9, 3-hydroxy-butyrate moved behind acetate despite its higher mobility ("enforced" isotachophoresis), but the sequence of their zones was reversed at higher pH values. The effect of the pH on the migration rate is illustrated by the slight decrease of the zone lengths in the recorder tracings (Fig. 2a-c) at higher pH values. The effect of pH on the step height can be used in the identification of ionic species. Sometimes it is easier to run a sample at different pH values rather than to develop an electrolyte system that allows separation of all components.

Quantitative determinations are generally performed using calibration curves. Fig. 3 shows the calibration curve for lactic acid at pH 4.0. As outlined above, it is necessary to construct a separate calibration curve for each pH value. Since the concentration in

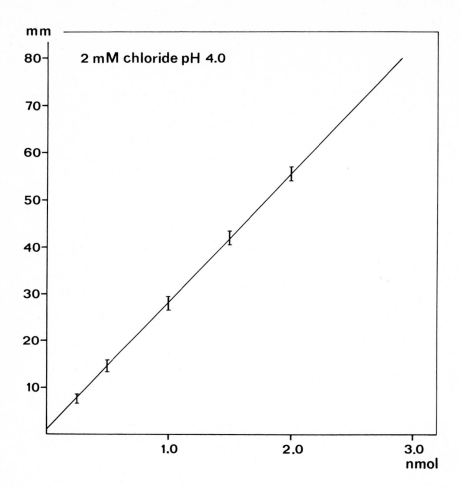

Fig. 3. Calibration line for lactic acid. Zone length in mm against
nanomoles injected. Analyses were performed in 2.5 mM chloride
as leading ion at pH 4.0.

consecutive zones following the leading ion decreases it might be
necessary to construct a calibration curve for each component in the
sample. In many cases the slopes of these calibration curves will
not be significantly different due to the error in the determinations.
The reproducibility of the method is satisfactory (approximately 5 %)
and it is not necessary to use internal standards.

The isotachophoretic analysis at pH 3.9 of blood of a healthy
individual is shown in Fig. 4. Lactic acid and a small amount of

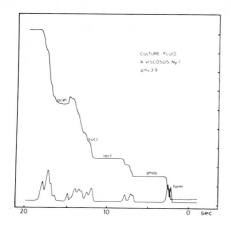
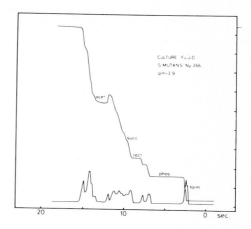

Fig. 4. Isotachopherogram of blood (a) 0.5 µl injected directly into
the separation capillary, and (b) blood plus standard solution
containing phosphate, acetoacetate, lactate and 3-hydroxybu-
tyrate. Leading electrolyte and recorder speed similar to
Fig. 1.

3-hydroxybutyric acid could be detected. A linear relationship was
found between isotachophoretic and enzymatic (lactate dehydrogenase)
analyses of lactic acid in blood (Fig. 5). Under the experimental
conditions used the concentrations of acetoacetate and 3-hydroxybuty-
rate are below the level of quantitative estimation. Due to the relatively high chloride content of the blood no samples larger than
approximately 0.5 µl could be injected. This difficulty could be
partially overcome by using a longer capillary, which has a larger
separation capacity. In order to reduce the time of the analysis, the
current could then be temporarily increased until the sample ions
have arrived near the detector.

Fig. 6 shows the presence of formic, lactic, and acetic acids in
human saliva. The acids are partly derived from bacteria in saliva·
The amount of acid is dependent on the number of bacteria in the saliva
and reflects the state of oral hygiene.

Fig. 7 shows the analysis of the fermentation products formic,
lactic, succinic and acetic acids in the culture fluids of Actinomyces

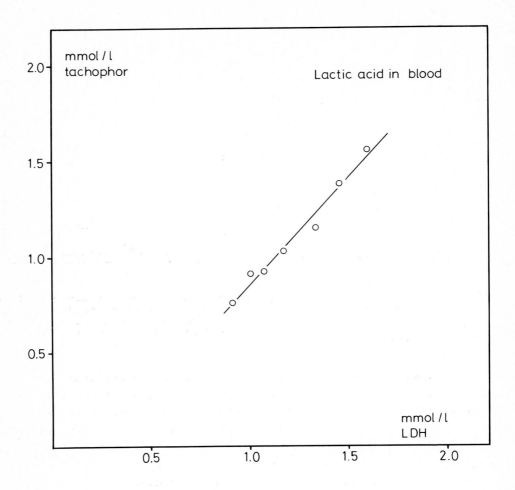

Fig. 5. Relationship between the enzymatic and the isotachophoretic
analysis of lactate in blood.

viscosus Ny1 and Streptococcus mutans Ny266 grown in a complex medium,
Trypticase soy broth. The amount of fermentable carbohydrate in this
medium is relatively small (approximately 15 mmoles of hexose per
liter) and the concentration of fermentation products is consequently
low.

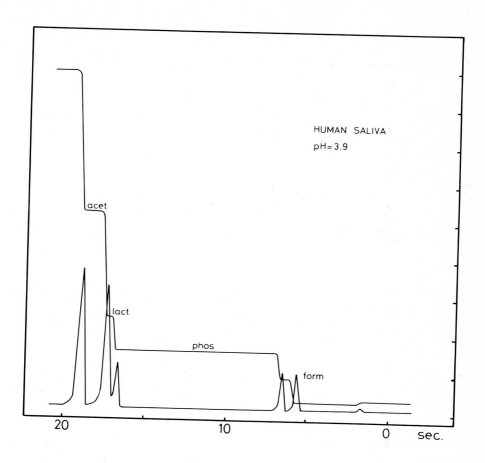

Fig. 6. Isotachopherogram of human saliva, 0.5 µl injected. Components
identified are indicated in Fig. 1. Leading electrolyte and
recorder speed similar to Fig. 1.

The isotachophoretic analysis of rat dental plaque is shown in
Fig. 8. The rat harbours an oral microflora containing various types
of bacteria such as actinomycetes and streptococci as well as Gram-
negative cocci and rods. A variety of anions was found to be present
in the plaque. Several organic acids arising from bacterial fermentation in dental plaque could be identified. The UV signal revealed
the presence of a small amount of an unidentified component that was

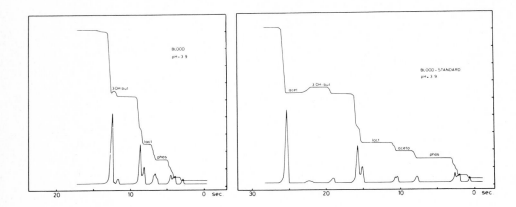

Fig. 7. Isotachopherogram of culture fluids of bacteria grown in
complex medium (Trypticase soy broth). (a) Actinomyces viscosus
Ny1, (b) Streptococcus mutans Ny266, 0.5 μℓ injected. Components identified are indicated as in Fig. 1. Leading electrolyte and recorder speed similar to Fig. 1.

not detected by the conductivity detector (Fig. 8). The isotachopherogram in Fig. 8, recorded at 500 mm/min, underlines the good resolution
of the conductivity cell.

The results show and imply that isotachophoresis is a suitable
method for the determination of a variety of organic acids in biological materials. The advantage of the method is that samples generally
do not require pre-treatment. The method is particularly useful for
the analysis of samples that are only available in small quantities.

The assay of small amounts of acid in the presence of a large excess
of a fast moving ion, such as chloride, is difficult and requires
special measures. For instance, 3-hydroxybutyrate can only be determined in blood if a long capillary is used and more sample is injected.

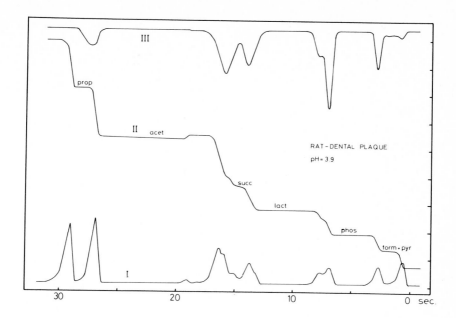

Fig. 8. Isotachopherogram of dental plaque from rats harbouring a
complex oral microflora. Leading ion 2.5 mM chloride, pH 3.9.
I, conductivity detector signal; II, differential signal,
III, UV detector signal. Recorder 500 mm/min.

The use of a conductivity detector for the identification and
quantitative estimation of ionic species was found to be fruitful.
We would recommend that a fast-reacting, sensitive step height detector
such as the conductivity cell, be provided by the manufacturer of the
Tachophor.

REFERENCES

1 F.M. Everaerts, J.L. Beckers and T.P.E.M. Verheggen, Isotachopnoresis
   Theory, Instrumentation and Applications, Elsevier, Amsterdam, Oxford,
   New York, 1976.
2 J.S. van der Hoeven, H.C.M. Franken, P.J.M. Camp and C.W. Dellebarre,
   1978. App. Environ. Microbiol. 35(1978)17-23.

A. Adam and C. Schots (Editors), *Biochemical and Biological Applications of Isotachophoresis*   81
© 1980 Elsevier Scientific Publishing Company, Amsterdam — Printed in The Netherlands

QUANTITATIVE CHARACTERIZATION OF PHOSPHORUS-CONTAINING ANIONS BY
ANALYTICAL ISOTACHOPHORESIS

Ph. Gysen and G. Heynen

Laboratoire de Radio-Immunologie (Prof. P. Franchimont)

Tour de Pathologie, bâtiment B23

C.H.U.   Sart Tilman   4000   Liège

INTRODUCTION

   Currently available quantitative methods for diphosphonate compounds
use the detection of phosphate by heptamolybdate, which forms coloured
phosphomolybdate complexes (ref. 1). A specific physicochemical method
has not been extensively studied although preliminary data suggest
that HPLC might be useful for the purpose (ref. 2).

   Isotachophoresis seemed to us another potential method for the
specific quantitative measurement of diphosphonate compounds.

MATERIAL AND METHODS

   Capillary isotachophoresis is performed using the 21-cm length
capillary tube, with the LKB 2127 Tachophor apparatus, at 25°C.

   The leading electrolyte is chloride (pH 2.7), the terminating
electrolyte caproic acid (pH 4), and the counter ion asparagine (10 mM).
Migration takes place at 100 μA and detection is performed at 280 nm,
even though the compounds studied are non-absorbing at this wave length
(ref. 3).

   All reagents, of analytical grade, are recrystallized before use.

   The following phosphorus-containing anions were studied :
1-hydroxyethane-1,1-diphosphonate (EHDP) ; 3-amino-1-hydroxypropane-
1,1-diphosphonate (APD) ; 1,1-dichloromethane-1,1-diphosphonate (Cl_2MDP);
phosphate (P) and pyrophosphate (PP).

The isotachophoretic mobility of each compound was characterized by internal standard and thermal detection. Quantitation was carried out by measuring the length of the non-UV-absorbing zone, which is proportionnal to the amount of the specific anion injected. A reference curve has been established for each compound (see Fig. 1).

RESULTS

Characterization of the compounds

The isotachophoretic mobilities of the compounds studied are different and can be arranged in the following order :
Cl_2MDP, PP, EHDP, P and APD. Each compound is characterized by a thermal signal, which is specific, as shown in Table 1.
The UV-absorbing peaks are due to the presence of small amounts of UV-absorbing anionic compounds of intermediate isotachophoretic mobility (ref. 3).

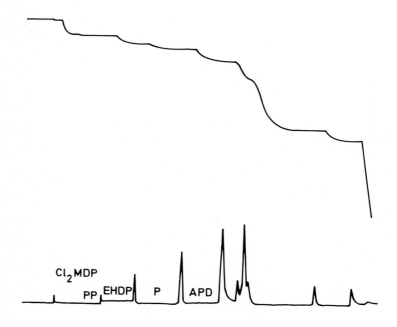

Fig. 1. UV (bottom) and thermal (top) profiles of a synthetic mixture of Phosphorus-Containing compounds by analytical isotachophoresis.

TABLE 1

Anion studied	Distance in mm between the level of the thermal line of the compound and the level of the leading, electrolyte
Cl_2MDP	7.5
PP	8.5
EHDP	13.0
P	15.5
APD	22.0

Quantitative measurements

Fig. 2 shows the linear relationship between the length of the non-UV-absorbing zone of EHDP and the amount of EHDP injected, allowing quantitative estimation. Sensitivity is dependent on two limiting factors : the minimum measurable length of the specific zone and the injection volume.

In practice, the length of the specific zone that can be measured with reliability is 2 mm, whereas the greatest volume that can be injected is 10 µl. Under our experimental conditions the sensitivity of the method is 1 nmole.

DISCUSSION

Isotachophoresis is a useful technique to quantitate and characterize anionic species (refs. 4 and 5).

The results show that PP, EHDP, Cl_2MDP, P and APD, all closely related compounds, can be effectively separated in a few minutes by isotachophoresis. The specificity of the technique has been established by use of an internal standard. The absolute sensitivity achieved is 1 nmole. Possible improvement of the sensitivity may be obtained by reducing the current to 50 µA;

CONCLUSION

As many other anionic species, disphophonate compounds can be quantitatively and specifically measured by analytical isotachophoresis.

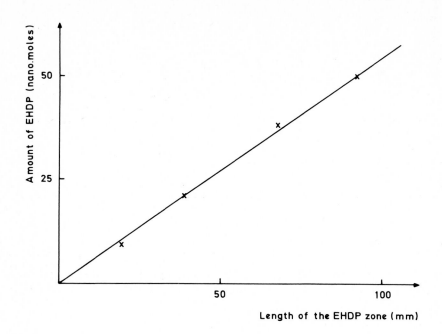

Fig. 2. Relationship between the amount of diphosphonate and the length of the UV-detection zone.

ACKNOWLEDGEMENTS

Diphosphonate compounts were kindly provided by Henkel GmbH, Düsseldorf, G.F.R.

REFERENCES

1 H. Goldenberg and A. Fernandez, Clin. Chem. 12(1976)871.
2 C.J. Preston, A. Challa, A. Noorwali, D.L. Douglas and R.G.G. Russel, Mineral Electrolyte Metab., 2(1979)259.
3 L. Arlinger and H. Lundin, Protides Biol. Fluids, Proc. Colloq., 21(1973)667.
4 F.M. Everearts, M. Geurts, F.E.P. Mikkers and Th.P.E.M. Verheggen, J. Chromatogr., 119(1976)129.
5 P. Boček, B. Kaplanova, M. Deml and J. Janák, J. Chromatogr.,153 (1978)287.

THE QUANTITATIVE ESTIMATION OF CALCIUM-DIPHOSPHONATE INTERACTION
BY ISOTACHOPHORESIS

Ph. Gysen and G. Heynen

Laboratoire de Radio-Immunologie (Prof. Franchimont)

Tour de Pathologie, bâtiment B23

C.H.U.   Sart Tilman   4000 Liège

INTRODUCTION

Various diphosphonates retard the growth and dissolution of hydro-
xyapatite crystals in vitro (refs. 1 and 2). Furthermore, they form
complexes with most divalent cations (Refs. 3 and 4), especially with
calcium. The metal is chelated in an electrostatic field and may form
a soluble mononuclear complex (ref. 4) following the scheme :

Soluble mononuclear
complex

Thus, there seems to be an equilibrium between free diphosphonate
and the soluble complex. Furthermore, the presence of a precipitate
allows the measurement of a solubility product.

Since we had a specific method for the quantitative measurement
of various diphosphonates (ref. 5), we have studied the quantitative
aspects of calcium-diphosphonate interaction in the presence of a
four-fold excess of calcium.

MATERIAL AND METHODS

A 1-ml volume of a 50 mM solution of the two diphosphonates, EHDP and APD, was added to 1 ml of a 200 mM CaCl solution. A precipitate formed immediately. Diphosphonate was measured in the supernatant and in the pricipitate, which was dissolved by the addition of 2 ml of 2 N HCl, by isotachophoresis as previously described (ref. 5).

In one experiment, the quantitative measurement made by isotacho-phoresis was compared to the results obtained by using ^{14}C-labelled APD (gift from Henkel GmbH).
The precipitate was resuspended in water and shaken for 5 min. APD was measured in the supernatant.

RESULTS

Table 1 shows that, in the presence of a four-fold excess of calcium 87 % of APD and 71 % of EHDP are precipitated at pH 5. The respective quantities of APD in the precipitate and the supernatant determined by isotachophoresis and by radioactivity, are the same. It should be noted that the recovery of APD by both methods is 100 %. There is no detectable APD in the supernatant of the precipitate resuspended in water.

TABLE 1

Repartition of APD and EHDP, both in the supernatant and the precipi-tate, measured by two methods

Diphosphonate	Method	% in the supernatant	% in the precipitate	pH
APD	Isotacho-phoresis	11.0	88.8	5
APD	Radio-acitvity	14.2	86.1	5
EHDP	Isotacho-phoresis	29	71	5

DISCUSSION

The present results show a good relationship between the recovery of labelled APD and the measurement of APD by isotachophoresis. In the presence of a four-fold excess of calcium, APD and EHDP form an aggregate which precipitates.

Within 5 minutes of the formation of the precipitate, we measured the quantities of APD and EHDP in both the supernatant and the precipitate. The result shows that precipitation was not complete. As the recovery of free and precipitated diphosphonate is 100 %, the presence of a soluble intermediate mononuclear complex in a significant concentration seems unlikely. However, it is possible that during isotachophoresis the soluble diphosphonate-calcium complex dissociates, and it would then be measured as free diphosphonate. The results of our experiment of washing the precipitated complex with water suggest that the precipitation is irreversible, as previously shown by Uchtman (ref. 3) for EHDP.

REFERENCES

1 M.D. Francis, Calc. Tiss. Res., 3(1969)151.
2 H. Fleisch, R.G.G. Russel and M.D. Francis, Science, 165(1969)1262.
3 V.A. Uchtman, J. Phys. Chem., 76(1972)1304.
4 Ph. Gysen and G. Heynen, unpublished results.
5 Ph. Gysen and G. Heynen, in Biochemical and Biological application of Isotachophoresis. A. Adam and C. Schots (Editors) Elsevier, Amsterdam, Oxford, New York, (1980), p. 81-84.

A. Adam and C. Schots (Editors), *Biochemical and Biological Applications of Isotachophoresis*
© 1980 Elsevier Scientific Publishing Company, Amsterdam — Printed in The Netherlands

ISOTACHOPHORETIC DETERMINATION OF 5-FLUOROURACIL IN SERUM

Bengt GUSTAVSSON

Department of Surgery II, Sahlgren Hospital, S-413 45 Gothenburg
(Sweden)

and

Astor BALDESTEN

Research Laboratories, LKB-Produkter AB, Fack, S-161 25 Bromma
(Sweden)

INTRODUCTION

5-Fluorouracil (5-FU) is one of the most commonly used drugs for
the palliative treatment of solid tumours of the breast and of the
gastrointestinal tract (ref. 1). Since its introduction some 20 years
ago various dose regimens and routes of administration have been used,
largely empirically (ref. 1). Since individual therapy is an important
concept of clinical pharmacology several analytical methods for the
determination of serum concentrations of the drug have been developed.
We now report an isotachophoretic assay of 5-FU in serum, which is
sensitive to 100 pmol/ml of plasma or serum.

METHODS AND PROCEDURES

The isotachophoretic separations were made in the LKB 2127 Tachophor.
Normal serum was obtained from healthy blood doners and kept at -20°C
until used for multiple analysis. Serum samples from patients receiving
250 mg of 5-FU orally were analysed by the isotachophoretic technique
as well as by the microbiological method (ref. 2). 5-FU was supplied
from Roche, Stockholm, Sweden. HPMC (hydroxy-propyl-methyl-cellulose)
Methocel 90 HG 15.000 CPS, was obtained from Dow Chemical Midland,
Mich., U.S.A. The resin AG 1-X8, 200-400 mesh in the formate form was
purchased from Bio-Rad Labs., Richmond, Calif., U.S.A. All chemicals
used were of analytical grade and commercially available if not otherwise stated.

Deproteinisation of serum

A 1-ml volume of plasma or serum was diluted to 5 ml with 1 % picric

acid. The precipitate formed was removed by centrifugation. The supernatant was brought to pH 11 by the addition of 0.1 N KOH and the sample
was then loaded onto an anion-exchange column.

Sample preparation.

The resin was equilibrated in 6 N formic acid for 1 h (1 part resin
to 5 parts of acid). The column dimensions are given in Fig.1. The
resin was washed with 30 column volumes of glass-distilled water.
Then the deproteinised serum sample was loaded onto the column. Twenty
column volumes of glass-distilled water (pH 7) were passed through
the column. 5-FU was eluted with 0.05 N formic acid. The eluate was
evapored to dryness at 60°C under a gentle stream of nitrogen, and
the dry sample was dissolved in 20 µl of water.

Isotachophoretic conditions

During the investigation different lenghts of capillary were used.
For routine purposes a 23-cm long capillary (I.D. 0.5 mm) proved
satisfactory. The separations were started at a current of 90 µA
which was reduced to 45 µA prior to detection.
The ligth transmission at 254 or 278 nm was registered on a recorder
at a chart speed of 2-5 cm/min.

Electrolytes

Leading : 5 nM HCl was titrated with 0.1 mM TRIS to the desired pH.
The leading electrolyte was 0.4 % in HPMC.

Terminating : 0.04 M glycine was titrated with satured $Ba(OH)_2$ to
the desired pH.

RESULTS AND DISCUSSION

5-FU could not be detected in serum directly injected into the
Tachophor. After the deproteinisation with picric acid and concentration
on the Ag 1-X8 column a distinct single zone with a mobility similar
to 5-FU in aqueous solution was obtained. However, it was necessary
to wash the column with at least 30 resine volumes of water to remove
heavily charged impurities from the resin which otherwise caused an
unduly long analysis time. These compounds are probably polymerised
sulphonated styrenes, as discussed by Greenlang et al (ref. 3). This
phenomenom seems to be of general importance in all situations where
biological samples are extracted on anion-exchange columns before the
isotachophoretic separation.

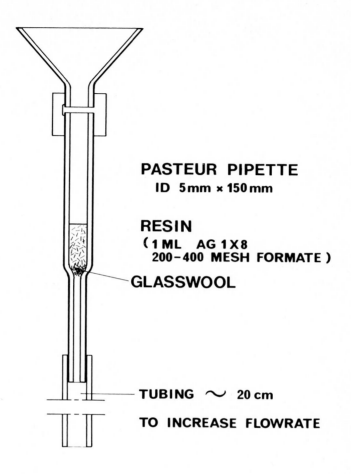

PASTEUR PIPETTE
ID 5mm × 150mm

RESIN
(1 ML AG 1X8
200-400 MESH FORMATE)
GLASSWOOL

TUBING ∿ 20 cm

TO INCREASE FLOWRATE

Fig. 1. Column used to extract 5-FU from serum. The funnel is attached
     to the top of the column with a piece of tubing.

   Different pH values in the leading and terminating electrolyte have
been investigated. The best resolution of the samples was obtained with
a pH of 7.45 in the leading electrolyte and 9.4 in the terminating
electrolyte.

   In spite of the deproteinisation and ion-exchange chromatography
there are several zones in the isotachopherogram of the 0.05 N formic

acid fraction. In order to identify 5-FU in such a separation it is
necessary to inject the sample alone and then together with the pure
drug. Another possibility is of course to analyse a blood sample from
the patient before he was given the drug and then compare the pattern
as shown in Fig. 2.

The methodological error from diplicate determinations was $\pm$ 6.2 %
when 0.1-10 nmol of 5-FU were analysed. The reproducibility of the
method was confirmed by running standard curves with increasing amounts
of 5-FU in aqueous solution and standard curves of 5-FU added to normal
serum. It was then found that the overall recovery of the method was
82 %. Moreover, the specificity of the method was evaluated by comparing a microbiological assay (ref. 2) and the isotachophoretic assay.
A close correlation was obtained (r = 0.99) when the content of 5-FU
was determined by the two different methods (Fig. 3).

It may be possible to improve the isotachophoretic assay, by the
production of an appropriate internal standard. Log-linear conversions
and integration of the areas under the UV-absorbance curves might increase the sensitivity of the method (ref. 4). Previous analytical
methods include radioactive labelling, gas chromatographic assays or
various gas chromatographic-mass spectrometric methods. These assays are
either not sensitive enough, or are time-consuming (ref. 5), expensive
(ref. 6) or too complex (ref. 7) for routine hospital use. Recently
high-pressure liquid-chromatographic assays of5-FU have been published
(refs 5 and 8). In conformity with the isotachophoretic assay these
methods have the potential not only to measure the parent drug 5-FU
but also its metabolites.

Although the aim of developing the isotachophoretic assay was to
measure 5-FU, several other compounds in serum are quantitatively
separated. Hypoxanthine, a metabolite of nucleotide metabolism, can
be measured by a modification of the method (ref. 9). Moreover, the
extraction technique and the isotachophoretic separation can be readily
adapted to the study of 5-FU in other biological material, such as
tumour tissue.

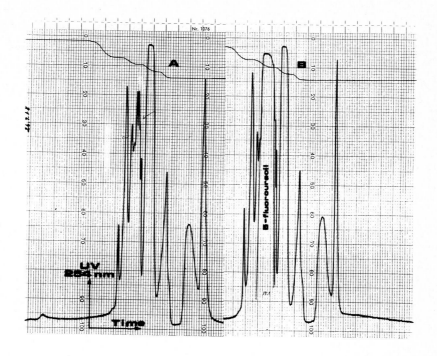

Fig. 2. A, Isotachopherogram of serum (prepared as described in the text) from a patient before 5-FU was administered. B, Isotachopherogram from the same patient after administration of 5-FU. The zone representing 5-FU is indicated.

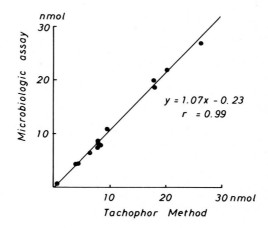

Fig. 3. Comparison of analytical methods for determination of 5-FU.

In summary we have found that 5-FU can be identified and quantitated in serum by analytical isotachophoresis. The technique is simple and the sensitivity at least the same as more complex gas chromatographic-mass spectrometric methods. This new assay can be used for the routine control of the serum levels of 5-FU after administration of this drug to patients. The extraction technique and the isotachophoretic separation should be readily adaptable to the study of 5-FU and its metabolites in tumours and normal tissues.

ACKNOWLEDGEMENT

This research was supported by the Swedish Cancer Society.

REFERENCES

1 W. Sadee and C.G. Wong, Clin. Pharm., 2(1977)437-450.
2 O. Almersjö, Acta Path. Microbiol. Scand., 85 N° 3(1977)227-234.
3 R.D. Greenland, Anal. Biochem., 62(1974)305-309.
4 D.C. Gower and R.C. Woledge, Sci. Tools, 24(1977)17-21.
5 J.L. Cohen and R.E. Brown, J. Chromatogr., 151(1978)237-240.
6 C. Finn and W. Sadee, Cancer Chemother. Rep., 59,N° 2(1975)279-286.
7 B.L. Hillcoat et al., Brit. J. Clin. Pharmacol., 3(1976)135-143.
8 N. Christophidis, Clin. Chem.,25,N° 1(1979)83-86.
9 B. Gustavsson, A new assay of hypoxanthine in serum by isotachophoresis (in preparation).

A. Adam and C. Schots (Editors), *Biochemical and Biological Applications of Isotachophoresis*   95
© 1980 Elsevier Scientific Publishing Company, Amsterdam — Printed in The Netherlands

PURITY CONTROL OF SYNTHETIC NAD[+] DERIVATIVES BY ISOTACHOPHORESIS

Jean Buret and Claude Cox

Laboratoire de Chimie médicale, (Prof. C. Heusghem), Rue des Bonnes-Villes 1, B4020 LIEGE (Belgium)

The coupling of NAD[+] to other molecules is useful in preparing special enzyme systems. The first step of the synthesis consists of the addition of short aliphatic chains to the NAD[+] (Fig. 1). To preserve the coenzyme properties of the NAD[+] derivatives, only the adenine ring can be modified.

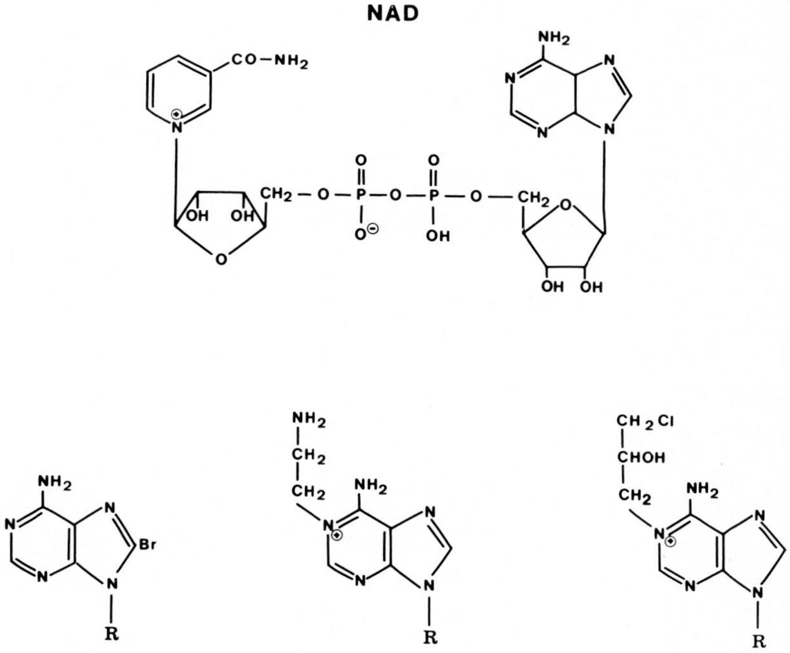

Fig. 1. Structure of NAD[+] derivatives

For this study, we have prepared three NAD^+ analogues : 8-bromo-NAD^+; N^1-(2-aminoethyl)-NAD^+; and N^1-(3-chloro-2-hydroxy-propyl)-NAD^+.

8-Bromo-NAD^+

This compound is synthesised by adding bromine water to NAD^+ at pH 4 (ref. 1). Purification of 8-bromo-NAD^+ is performed by acetone and ethanol precipitations and by ion-exchange chromatography [Dowex 1-X2 (Cl^-) and Dowex 1-X2 ($HCOO^-$)].

N^1-(2-aminoethyl)-NAD^+

The aminoethyl side-chain is attached by the addition of aziridine to NAD^+. The pH of the solution is maintained automatically at 4 for 60 h (refs. 2 and 3). To purify the derivative, we used acetone pre-cipitation and ion-exchange chromatography on DEAE-Sepharose followed by desalting on Sephadex G-10.

N^1-(3-chloro-2-hydroxy-propyl)-NAD^+

Epichlorhydrine is added to NAD^+ at pH 6 which is maintained cons-tant for 60 h (ref. 4). The first steps of the purification of this derivative are the same as for N^1-(2-aminoethyl)-NAD^+.

These three products are only intermediates in a longer synthesis. Nevertheless, it is interesting to check their purity.

The tests for purity of the products obtained (frequently con-taminated by mineral salts) must be able to detect the presence of degradation products of NAD^+ or other NAD^+ derivatives. Few techniques are suitable : UV spectrophotometry; thin-layer chromatography; and measurement of coenzyme activity. This is why isotachophoresis was chosen to control the purity of these analogues. However, as we had a limited goal, we were only interested in the UV recordings showing the degradation products of NAD^+ and the side-products. All the mineral salts are recorded by the thermal detector. As these records do not give useful information for our purpose, they are not given in the figures.

MATERIAL AND METHODS

The LKB 2127 Tachophor is equipped with a UV detector set at 280 nm. The temperature is kept constant at 25°C. All reagents were purchased from Aldrich Europe (Beerse, Belgium). Hydroxy-propylmethylcellulose (HPMC) was generously given by Dow Chemical (Brussels, Belgium).

RESULTS

Isotachophoresis of 8-bromo-NAD^+

The pK of this product is about 4. Therefore, we use an acidic system.

Leading electrolyte : HCl 5 mM
                         β-alanine 18 mM (pH = 4.00)
                         HPMC K 15 M 4 g/l.
Terminator : isobutyric acid 5 mM (pH = 3.63).
Capillary length : 63 cm.
Electric current : 70 µA.

The recording (Fig. 2) shows the purity of the 8-bromo-NAD^+ which is contaminated by UV-absorbing materials. There are also some salts. It is also possible (Fig. 3) to separate NAD^+ and 8-bromo-NAD^+. Comparison of Figs. 2 and 3 shows that 8-bromo-NAD^+ is free from NAD^+.

Isotachophoresis of N^1-(2-aminoethyl)-NAD^+

We were unable to perform isotachophoresis of this compound in an acidic system. This probably due to the formation of an internal ion between one phosphate and the amines. To give a negative charge to the molecule, it is necessary to run the isotachophoresis in an alkaline medium (refs. 3 and 5).

Leading electrolyte : 4-morpholine-ethanesulfonic acid (MES) 5 mM
                         2-amino-2-ethyl-1,3-propanediol (AMPD) 10 mM
                         HPMC K 15 M 4 g/l.
Terminator : 6-aminocaproic acid (EACA) 11.4 mM
                 AMPD 10 mM
                 barium hydroxide saturated solution to pH 10.8.
Capillary length : 23 cm.
Electric current : 75 µA.

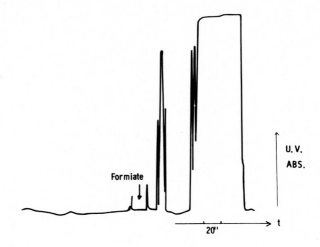

ISOTACHOPHORESIS OF 8-BR-NAD$^+$: 7 μGR
LEADING : HCL 5 MM
β-ALANINE 18 MM } PH = 4.00
TERMINATOR : ISOBUTYRIC ACID 5 MM (PH = 3.63)
CAPILLARY LENGTH : 63 CM
ELECTRIC CURRENT : 70 μA

Formiate

U.V.
ABS.

20'

t

Fig. 2.

Except for the presence of salts (Fig. 4), the degree of purity is good enough to utilize this product for other syntheses.

Isotachophoresis of N^1-(3-chloro-2-hydroxy-propyl)-NAD$^+$

The isotachophoresis of this analogue was run in the alkaline system described for the N^1-(2-aminoethyl)-NAD$^+$. The recording (Fig. 5) shows that the purification steps are inadequate : the first UV-absorbing zone has an area of about 30 % of the NAD$^+$ derivative.

CONCLUSION

Because of their lack of specificity the usual techniques of ana-lytical chemistry are useless for molecules such NAD$^+$.
Fortunately, isotachophoresis has sufficient specificity that it can separate these molecules as they differ in their effective mobilities.

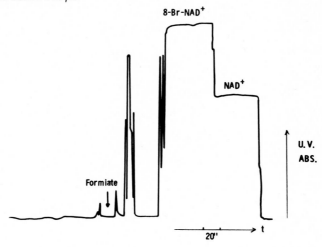

Isotachophoresis of NAD$^+$ and 8-Br-NAD$^+$: 7 μgr and 7 μgr

Leading : HCl 5 mM
β-alanine 18 mM $\Big\}$ pH = 4.00

Terminator : isobutyric acid 5 mM (pH = 3.63)

Capillary length : 63 cm

Electric current : 70 μA

8-Br-NAD$^+$

NAD$^+$

U.V. ABS.

Formiate

20'

t

Fig. 3.

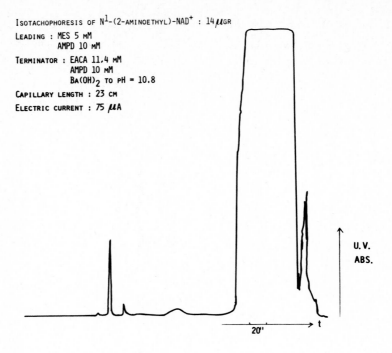

ISOTACHOPHORESIS OF N^1-(2-AMINOETHYL)-NAD$^+$: 14 μGR

LEADING : MES 5 mM
AMPD 10 mM

TERMINATOR : EACA 11.4 mM
AMPD 10 mM
Ba(OH)$_2$ TO pH = 10.8

CAPILLARY LENGTH : 23 CM

ELECTRIC CURRENT : 75 μA

U.V.
ABS.

20'

t

Fig. 4.

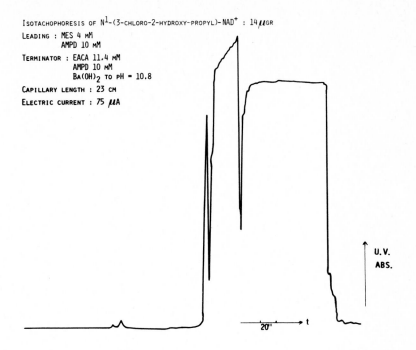

Isotachophoresis of N^1-(3-chloro-2-hydroxy-propyl)-NAD^+ : 14 μgr

Leading : MES 4 mM
            AMPD 10 mM
Terminator : EACA 11.4 mM
             AMPD 10 mM
             Ba(OH)$_2$ to pH = 10.8
Capillary length : 23 cm
Electric current : 75 μA

U.V.
ABS.

20'

t

Fig. 5.

The sensitivity of this analytical tool is also an advantage.
It is only necessary to take, after each step in a synthesis, the
few micrograms needed for the isotachophoretic analysis.

REFERENCES

1 M.A. Abdallah, J.F. Biellmann, B. Nordström and C.I. Bränden,
  Eur. J. Biochem., 50, (1975),475-481.
2 H.L. Schmidt and G. Grenner, Eur. J. Biochem. 67(1976)295-203.
3 H.G. Windmueller,and N.O. Kaplan, J. Biol. Chem., 236(1961)2716-
  2726.
4 G. Grenner, H.L. Schmidt and W. Völkl, Hoppe-Seyler's Z. Physiol.
  Chem., 357(1976)887-891.
5 M.K. Weibel, C.W. Fuller and J.M. Stadel, Enzyme Eng., 2(1974)203-208.

A. Adam and C. Schots (Editors), *Biochemical and Biological Applications of Isotachophoresis*   103
© 1980 Elsevier Scientific Publishing Company, Amsterdam — Printed in The Netherlands

ISOTACHOPHORETIC ANALYSIS OF METABOLITES IN MUSCLE EXTRACTS

R.C. Woledge and P. Reilly

Department of Physiology, University College, London W.C. 1
(Great Britain)

INTRODUCTION

Analysis of muscle extracts in our laboratory using isotachophoresis has been described previously (ref. 1). Both UV-absorbing materials (e.g. ATP and ADP) and non-UV-absorbing materials (e.g. PCr) can be measured. We have recently modified the conditions to improve the resolution and sensitivity. Also we have replaced the manual method of measurement of the records by electronic integration of logarithmic transformations of the transmission signal. This paper describes these improvements.

METHODS

Muscle extraction

As in our earlier work frog muscles at 0°C were frozen by the hammer freezing apparatus (ref. 2). Each frozen muscle (30-60 mg wet weight) was placed in 3 ml of 1.25 mM EDTA (pH 7.6) in 50 % methanol at -30°C. After four days the muscles were removed, freeze-dried and weighed. The methanol extract was freeze-dried and redissolved in 3 ml of water. Isotachophoretic analyses were performed on 10 $\mu\ell$ samples of this extract. It is also possible to analyse muscles extracted in other buffers or $HClO_4$, but the separation is slow when the extract contains a higher salt concentration.

Electrolytes

The leading electrolyte was 2.5 mM HCl with 7.7 mM β-alanine as a counter ion, in 2 % methylcellulose (pH was not measured but was probably about 3.9). The terminating electrolyte was 2.5 mM n-caproic acid. Methylcellulose was used to increase the viscosity of the medium and thus minimise the effects of electroendosmosis, which otherwise limits the resolution

Analyses were performed with the LKB Tachophor using a single-turn capillary. Perspex supports for the syringes which store and inject the buffers were constructed. A screw was used to push the plungers of the syringes making the use of more viscous solutions feasible, and probably improving the repeatibility with which the junction between the two buffers could be formed.

The addition of methylcellulose to the leading electrolyte introduced contaminating substances. However, most of this material could be removed by centrifugation and by purification of the methylcellulose stock solution on an ion-exchange column (BDH Amberlite). Stock solutions of 1 N HCl were made from concentrated HCl, because use of mercury-containing volumetric solutions resulted in the formation of a large UV-absorbing zone, probably of a Hg-EDTA complex.

Records and integration of peaks

Separations, which required about 20 min each, were started with a current of 100 µA, which was reduced to 25 µA shortly before detection of the separated zones by their UV absorption at 254 nm. Recorder paper speed of 10 cm/min was used. The recorder paper drive and the computing integrator were switched on by an automatic triggering system connected to the thermal detector of the Tachophor. Since the thermal detector is "upstream" from the UV detector the paper was switched on slightly before UV detection of the first zone. The thermal detector was not used for any other purpose.
The Infrotronics computing integrator was programmed to read the signal from the UV detector at 0.1-sec intervals throughout the run.

Maintenance

The capillary is periodically cleaned by using suction to pull through 200 ml of 1 % Triton X-100, followed by 200 ml of 0.1 N HCl, and 2000 ml of water. The cellophane membrane and the injection port septum are also changed at this time. At the end of each day's use, the capillary and reservoirs were rinsed and filled with water. Leaving water in the capillary helped to eliminate the problem of bubbles. forming during runs. The UV light remained on continuously, with the shutter closed when not in use, and the UV signal was standardised daily. Electrolytes were made up fresh each week and stored at 4°C.

RESULTS

Composition of electrolytes

The HCl concentration in the leading electrolyte was reduced, from the previously used 5 mM, to 2.5 mM. This reduced the O.D. of the UV-absorbing zones and therefore gave a better signal-to-noise ratio. With the lower concentration of HCl separation could be achieved in the same time with less current. We feel that this has been responsible for the absence of any bubble formation during the runs, which has sometimes been a problem in the past. Probably, at low HCl concentrations more methylcellulose is required to give a good resolution. Fig. 1 shows the effects of increasing the concentration of methylcellulose in the leading electrolytes from 0.5 % to 2 %. We also tried using 3 and 4 % methylcellulose, but these solutions gave no better resolution than 2 % and were difficult to use, because of their high viscosity. A concentration of 2 % methylcellulose was therefore adopted.

Analysis of muscle extracts

Fig. 2 shows a typical record of a muscle extract, with an insert showing the "blank" run in the absence of muscle extract. The zones are identified as to their content. Records were made with different amounts of this extract, from 0.5 µℓ to 10 µℓ, to find out how the zones in the blank are distributed among the zones formed by the muscle extract. The lower-case letters in the figure show the conclusion reached. The UV-absorbing impurity that appears between the Pi zone and the PCr zone is a convenient marker allowing these zones to be separately measured.

Table 1 shows repeatability of the results obtained in the analysis of several muscle extracts. The coefficient of variation is greatly reduced after normalisation of zone area, or zone length, by the measured zone length for EDTA. This indicates that the major source of variability is in the effective amount of material injected. It seems that EDTA can serve a useful purpose as an internal standard if carefully measured at the time of extraction. Repeatability of the ATP, Pi and PCr zones is about 1.5 % to 2 % after normalisation with EDTA zone length. The ADP zone is more variable, probably because of its relatively small size, or failure to separate completely from some blank component.

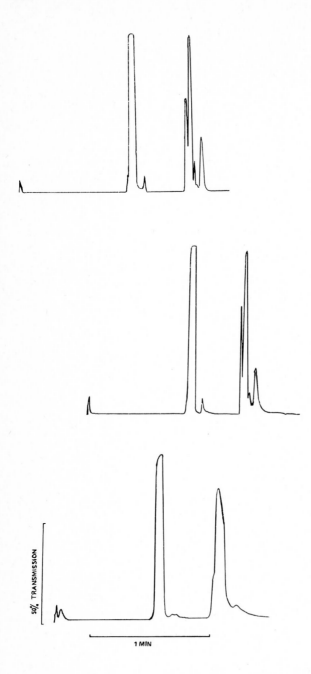

Fig. 1. The effect of different methylcellulose concentrations in the leading electrolyte. Three analyses of the same muscle extract. Upper record 2 %, middle record 1 % and lower record 0.5 % methylcellulose.

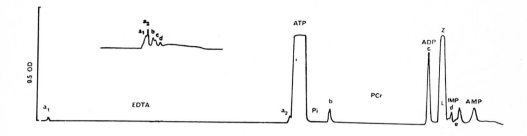

Fig. 2. Identification of zones in analysis of a muscle extract.
The lower-case letters mark the zones present in the blank,
which is shown as an inset. The zone marked Z is probably an
EDTA-heavy metal complex.

TABLE 1

Coefficients of variation for repeated analysis of various metabolites
by isotachophoresis.

All analyses were made with 10 µℓ samples. ATP and ADP results are
from the integrated areas under the peaks. The PCr, EDTA and Pi results
are from the measurement of zone lengths. Each result is from 15
measurements.

Metabolite	Standard deviation / mean (%)	
	Not normalised	Normalised by EDTA zone
ATP	4.1	1.9
ADP	4.4	5.1
PCr	3.9	1.8
EDTA	3.6	–
Pi	3.7	1.7

Volumes of sample greater than 10 µℓ tended to overload the system
causing systematically low values for all the materials measured.
However it is possible to use a more concentrated extract, since we
have obtained excellent separation of ATP, for example, at levels as
high as ten times the normal amounts seen with 10 µℓ samples of these
muscles, provided the volume injected did not exceed 10 µℓ.

Calibrations were carried out by injections of standard solutions. A linear calibration curve over a range of 0.1 to 10 nmol was obtained for ATP, from 0.01 to 0.25 nmol for ADP, from 0.5 to 5 nmol for PCr and from 0.1 to 10 nmol for Pi. These ranges are not necessarily the upper and lower limits of resolution of the systemn as we were interested in standardisation over the range of concentrations normally encountered with our muscle extracts..

Comparison of ATP and PCr concentrations determined by isotachophoresis with the results of other methods of measurement shows good correlation, as reported in a previous paper. In the case of Pi, however, the zone length seen with muscle extract is about twice that expected from enzymatic analysis of the same muscle extracts. This suggests that the Pi zone contains another component, derived from the muscle, which has about the same mobility as Pi. We have not been able to find conditions which will cause these two materials to separate.

DISCUSSION

Isotachophoresis provides a simple means of analyzing muscle extracts. The advantages are that analysis requires very small samples, and yields many observations from a single run, providing a direct comparison of the relative amounts of different metabolites. Thus it is convenient to use this method when many analyses are required on relatively few samples. If many samples are to be analyzed, it is less convenient than other methods, since each run takes about 20 min. The sensitivity using isotachophoresis is low for non-UV-absorbing materials and moderate for UV-absorbing ones. Mixed zones can form, making it difficult to measure some materials such as Pi. In combination with other methods isotachophoresis has proved a valuable tool in research into muscle energetics.

REFERENCES

1 D.C. Gower and R.C. Woledge, Sci. Tools, 24(1977)17-21.
2 K.M. Kretzschmar and D.R. Wilkie, J. Physiol., 202(1969)66P.

A. Adam and C. Schots (Editors), *Biochemical and Biological Applications of Isotachophoresis*   109
© 1980 Elsevier Scientific Publishing Company, Amsterdam — Printed in The Netherlands

DETERMINATION OF URINARY OXALATE CONCENTRATIONS BY ANALYTICAL
ISOTACHOPHORESIS

K. Schmidt, V. Hagmaier and G. Bruchelt
Department of Surgery, University of Tübingen, Tübingen (G.F.R.) and
Division of Urology, Department of Surgery, University of Basle,
Basle (Switzerland).

SUMMARY

A large number of different procedures have been described in
the literature for the quantitative analysis of urinary oxalate
concentrations. However, a rapid and sufficiently simple technique
for routine determinations is lacking. In this study the application
of analytical isotachophoresis to the measurement of urinary oxalate
is described.
Different pretreatment procedures of the urine specimens such as
precipitation, use of internal standards, and the addition of spacer
electrolytes are compared. Within a short analysis time complete
separation of the oxalate anion from accompanying ions is achieved in
the isotachophoretic system. The extremely high resolving power and
accuracy of this technique in addition to the short analysis time
are of significant importance in studying the role of oxalate in
urinary stone disease.

INTRODUCTION

Recent epidemiological studies have demonstrated a 5 % frequency
of urinary tract stones in the population of western countries. Calcium
oxalate stones are the most common type of renal calculus, accounting
for about 70 % of all stones. Although calcium metabolism and excretion
have been investigated extensively in patients with recurrent nephrolithiasis, relatively little attention has been paid to oxalate.
New interest in oxalic acid has mainly been caused by the important role
of oxalate in controlling stone formation in patients with recurrent
nephrolithiasis and the determination of urinary oxalate has been
considered of some value in the elucidation of causes of this condition.

Exogenous intake and in particular endogenously formed oxalate are
considered important factors in stone formation. Glyoxylate, several
amino acids, and ascorbic acid have been clearly established to be
the main precursors of oxalate in man. Hodgkinson (ref. 1) has reviewed
a number of reports on the source of the oxalate excreted in the urine,
concluding that the intestinal absorption of oxalate is very low.
However, in some subjects with increased urinary oxalate excretion the
intestinal absorption rate seems to be drastically changed due to a
deficiency in oxalate-precipitating compounds. Increased binding of
calcium to fatty acids in the malabsorption syndrome results in increased absorption and renal excretion of oxalate.

For the determination of oxalate a number of different methods are
described in the literature. Volumetric procedures such as permanganate
titration (ref. 2), or cerate oxidimetry (ref. 3), were found to be
unspecific. Colorimetric (ref. 4) and fluorimetric methods (ref. 5)
are mostly of limited specificity due to the fairly critical quantitative reduction of oxalic acid to glyoxylic acid. Methods involving
enzyme reactions utilized oxalate oxidase (EC 1.2.3.4) or oxalate
decarboxylase (4.1.1.2) (ref. 6). The reactions are highly specific
for oxalate but have the disadvantage that the enzymes are difficult
to isolate and catalytic activity is inhibited by various urinary constituents such as phosphate or sulphate (ref. 7), so that only incomplete
cleavage of oxalate is achieved. Additionally gas-liquid chromatographic
methods and isotope dilution techniques are available.

Analytical isotachophoresis has recently been successfully applied
to the analysis of organic acids extracted from human urine. This paper
demonstrates the separation and quantification of oxalate in urine by
means of this simple, rapid,and sensitive technique.

MATERIALS AND METHODS

For the determinations, urine from ten young, healthy, male volun-
teers was collected during 24 h and kept frozen until analyzed. Subjects
were on a free diet but were instructed to avoid excess intake of food
containing large amounts of oxalic acid. Isotachophoretic runs were
performed in a Tachophor 2127 apparatus (LKB, Sweden) equipped with an
800 mm PTFE capillary of 0.5 mm internal diameter. Design and function
of the instrument are reviewed by Arlinger (ref. 8) and Everaerts et al.
(ref. 9).

The leading electrolyte was 0.01 M hydrochloric acid titrated to pH 3.68 with 0.01 M beta-alanine. O.3 % Methylcellulose (methocell MC 4000 cP) was added to increase viscosity and thus to reduce electro-endosmosis. The terminating electrolyte was 0.01 M n-hexanoic acid. The initial current was 175 µA, and the analyzing current 75 µA. The temperature was kept constant at 15°C. For injection of aliquots of the urine samples (10 µℓ injection volume) high-quality syringes with a PTFE-tipped piston were used to avoid iron-oxalate complex formation. For quantification of separated ions the width of the UV peaks (254 nm) corresponding to the zone length in the capillary was used.

Three different preatreatment schemes for the urine specimens were employed prior to the isotachophoretic analyses:
  (a) acidification with hydrochloric acid alone,
  (b) acidification and addition of a spacer electrolyte gradient (Ampholyte),
  (c) acidification, adjustment of pH to 5.3, addition of calcium chloride solution, centrifugation, and dissolution of the precipitate.

Identification and calibration were performed by use of oxalate and ^{14}C oxalate as internal standards as well as the thermosignal.

All chemicals and reagents were of highest commercially available purity; water was twice distilled in an all-quartz apparatus. Buffers were prepared fresh weekly.

RESULTS AND DISCUSSION

A typical result of the isotachophoretic analysis of suprapure oxalic acid dissolved in water is given in Fig. 1 in terms of the UV signal and the thermosignal, the thermal step height representing the temperature of the corresponding zone produced by Joules heat in the capillary. The thermal step height of oxalate amounts to about 8 % of the total thermosignal chloride-caproate. The UV recordings exhibit the weak but characteristic absorbance of the oxalate ion. Organic acids metabolically related to oxalic acid are well separated and exhibit a markedly different thermal step height (Fig. 2).

The cyclic parabanic acid, which decomposes easily via oxaluric acid to oxalic acid and urea, thus possibly causing errors in the

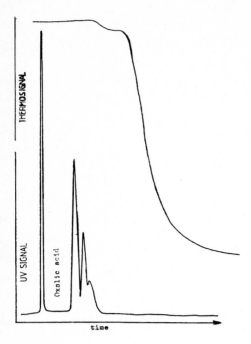

Fig. 1. Oxalic acid (suprapure) dissolved in water.

oxalate measurements, gives no signal because of its non-ionic
properties. However, cleavage of the ring and generation of oxalic
acid can easily be followed by isotachophoresis (Fig. 3 a and b).
Injection of urine into the Tachophor without preseparation of oxalic
acid results in a very complicated pattern of ionic constituents,
oxalate being one of the ions of highest mobility. Injection of urine
without pretreatment poses some significant disadvantages concerning
the evaluation of the tachopherograms and the analysis time.
In urine the total amount of oxalate is distributed to different pools:
oxalate anion, undissociated or crystallized oxalate salts, tetra-
oxalate, oxalato-complexes of transition elements. So variyng pool
sizes of oxalate can cause a reduced accuracy of the measurement if
not all the oxalate is converted into one homogeneous pool. Furthermore the small zone length of oxalate makes the graphic evaluation
difficult and the high ionic strength of urine results in a very
long analysis time.

 Addition of spacer electrolytes gives a better resolution of the
isotachophoretic separation but the quantitative evaluation of the

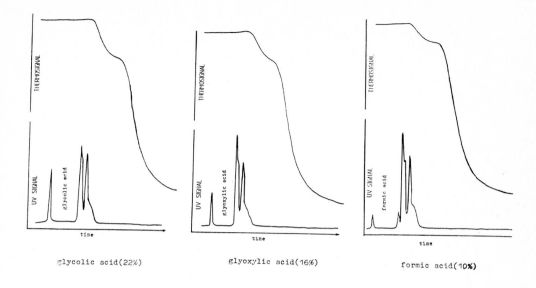

glycolic acid(22%)   glyoxylic acid(16%)   formic acid(10%)

Fig. 2. Isotachophoresis of organic acids metabolically related to oxalic acid.

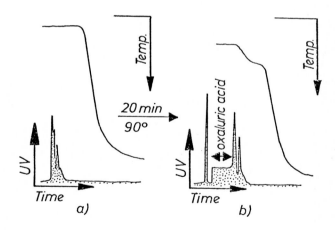

Fig. 3. Thermal decomposition of parabanic acid followed by analytical isotachophoresis

tachopherograms remains difficult because of the small zone length
of oxalate when normal urine is injected. Much better results are
obtained after precipitation of the oxalate with calcium ions and
injection of the dissolved precipitate into the Tachophor. Errors
caused by varying recovery rates are avoided using internal oxalate
standards.

Standard calibration curves of oxalate in urine show linearity
within a wide range of concentrations paralleling those curves obtained with oxalate dissolved in water (Fig. 4). The concentration
of the endogenous urinary oxalate can be measured on the negative
abscissa. The precision of the method as estimated from multiple
determinations of the same specimen was very high. A coefficient of
variation of 3.2 % was calculated. However, biological variation of
diurnal oxalate excretion was found to be very high. The mean value
in ten subjects was 29 $\pm$ 12 mg oxalate per day, which is comparable
with normal values determined by enzymatic methods.

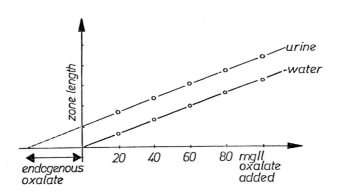

Fig. 4. Standard calibration curves of oxalate dissolved in water
or urine.

REFERENCES

1 A. Hodgkinson, Academic Press, London, 1977.
2 A.V. Harcourt, J. Chem. Soc., 20(1867)460.
3 G.H. Koch and F.M. Strong, Anal. Biochem., 27(1969)162.
4 S. Burrows, Analyst, 97(1972)958.

5 D.A. Britton and J.C. Guyon, Anal. Chim. Acta, 44(1969)397.
6 G.G. Mayer, D. Markow and F. Karp, Clin. Chem., 9(1963)334.
7 C.F. Knowles and A. Hodgkinson, Analyst, 97(1972)474.
8 L. Arlinger, Biol. Fluids, Proc. Colloq., 22(1975).
9 F.M. Everaerts, J.L. Beckers and Th.P.E.M. Verheggen, Isotachophoresis - Theory, Instrumentation and Applications. (J. Chromatogr. Libr., Vol. 6) Elsevier, Amsterdam, Oxford, New York, 1976.

A. Adam and C. Schots (Editors), *Biochemical and Biological Applications of Isotachophoresis*   117
© 1980 Elsevier Scientific Publishing Company, Amsterdam — Printed in The Netherlands

DETERMINATION OF URINARY OXALATE BY ISOTACHOPHORESIS IN UNTREATED
URINE

W. Tschöpe and R. Brenner
Medizinische Universitäts-klinik, Sektion Nephrologie,
Bergheimer Strasse 58, 69 Heidelberg (G.F.R.)

A. Baldesten
LKB Produkter AB, S-16125 Bromma 1 (Sweden)
and
E. Ritz
Medizinische Universitäts-Klinik, Sektion Nephrologie,
Bergheimer Strasse 58, 69 Heidelberg (G.F.R.)

SUMMARY

   The principle of isotachophoresis has been used to develop a simple,
rapid, specific and sensitive analytical method for the determination
of oxalate in untreated urine. The accuracy and precision of the method
were quantitated and found to be equal or superior to currently exis-
ting methods. Rapid and easy determination of oxalate in unprocessed
urine provides a useful tool for studies or renal stone disease and
oxalate metabolism.

INTRODUCTION

   The innumerable number of methods proposed for the determination
of urinary oxalate indicates that none of them is entirely satisfactory.
Most of the recently developed methods involve a two-step procedure:
separation of oxalate and its subsequent determination. For the isolation of oxalate, acid extraction by ether (ref. 1), by tri-n-butyl
phosphate (ref. 2), or precipitation (mostly as the calcium salt)
(ref. 3), or separation from interfering substances by ion-exchange
chromatography (ref. 4), or a combination of these procedures has been
used.
The following methods have been proposed for the determination of

oxalate in urine :
- oxalic acid decarboxylase (EC 4.1.1.2.) which specifically decar-
  boxylates oxalate to formic acid and CO_2 (refs. 5-7);
- colorimetric determination of oxalic acid with uranium (IV)-4-
  (2-pyridylazo) resorcinol (ref. 8);
- colorimetric determination with chromotropic acid after reduction
  of oxalic acid to glycolic acid (refs. 9 and 10).

It was the purpose of the present investigation to examine whether
oxalate in urine can be measured by isotachophoresis. Quantitation
of carboxylic acids by isotachophoresis has been demonstrated by
Beckers and Everaerts (ref. 11). Isotachophoresis for the determination
of urinary oxalate offers the combined advantages of simplicity and
avoidance of loss of oxalate during pretreatment procedures, since
isotachophoresis can be performed with untreated urine.

MATERIALS AND METHODS

Analytical system

Oxalate was determined in aqueous solution and urine by isotachophoretic separation, using an LKB 2127 Tachophor (LKB Compagny, Bromma,
Sweden) equipped with the 23 cm Teflon capillary (I.D. 0.5 mm). The
samples were injected, using 10 µl Hamilton syringes (serie 700) with
stainless steel cannulas, through the inlet membrane into the leading
electrolyte. The volume of the injected sample was adjusted to give approximately 1 nM oxalate; the average injected volume was 4 µl.

The leading buffer was HCl Suprapur (E. Merck, Darmstadt, G.F.R.)
5×10^{-3} M, NaCl (p.a. grade, E. Merck) 1×10^{-3} M, HPMC (Methocel
90 HG 15000 cps, Dow Chemical Co., Midland, Mich., U.S.A.). No adjustment to the pH was carried out. The terminating buffer was formic
acid 10^{-2} M (p.a. grade, E. Merck). Oxalate decarboxylase (EC 4.1.1.2.)
was purchased from Sigma Corporation, Munich, G.F.R. (N° 0-3500).

The current used was 150 µA, and the duration of the separation,
15-30 min. Detection was by spectophotometric absorption at 254 nm.
Chart speed was 1 mm (5 mm)/sec.
Fig. 1 gives a typical absorption curve for urinary oxalate. The
recording shows an initial UV-positive zone, followed by a non-UV-
absorbing zone. For the reasons given below, for standard curves and
samples, the zone lengths of the UV-positive and the UV-negative zones
were added.

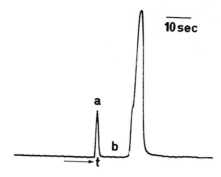

Fig. 1. Typical isotachophoretic run of a 24-h urine sample. Operational system as given above, chart speed 1 mm/sec. Only the
initial zones are shown, which are necessary for identification
of oxalic acid (a: initial UV-positive zone of oxalate; b: UV-
negative zone of oxalate). For a complete isotachophoretic
record of urine see Fig. 3. Oxalate concentration of the sample
was 545 μmoles/l, corresponding to a total of 2.18 nmoles
oxalate injected. The injected volume was 4 μl. Zone lengths
were measured with a scaled magnifier.

Collection of samples

Seventeen spontaneously voided urine specimens were collected
between 9.00 and 11.00 a.m. from healthy male individuals (age 21-55
years, mean 37 years). None of the probands (physicians) had a history
of renal stone disease or urinary tract infection and all had normal
renal function. Concentrated $ZnCl_2$ was then added (final concentration
10 mM). For the determination of recovery, increasing amounts of oxalate
were added to the urine as needed. Oxalate concentration was measured
by isotachophoresis within 12 h after the urine had been voided.

RESULTS AND DISCUSSION

Determination of oxalate in aqueous solution

As shown in Fig. 2a, spectophotometric measurements at 254 nm showed
a fast and small UV-absorbing zone, followed by a UV-negative zone .
Both the UV-positive and UV-negative zones increased in parallel upon
increasing the oxalate concentration. After treatment with oxalate

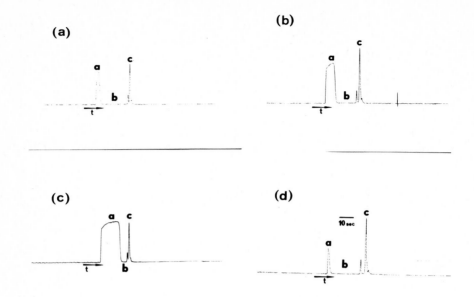

Fig. 2. The effect of adding ferric ions to aqueous solutions of oxalic
acid. (a) Injection of 4 nmoles oxalic acid. Operational system
as given above, chart speed 1 mm/sec. a,b: UV-positive and
UV-negative zones of oxalic acid; c: unknown impurities.
(b) Injection of 4 nmoles oxalic acid + 0.4 nmoles Fe^{3+}. The
initial UV-absorbing zone is noticeably increased. (c) Injection
of 4 nmoles oxalic acid + 0.8 nmoles Fe^{3+}: further broadening
of the UV-absorbing zone by ferric oxalate. (d) Injection of
4 nmoles oxalic acid + 0.8 nmoles Fe^{3+} + 4 nmoles EDTA:
disappearence of the UV-positive zone of ferric oxalate, most
oxalate now being recorded as UV-negative zone.

decarboxylase at pH 3, both the UV-positive and UV-negative zones
completely disappeared. It is concluded from this experiment that
the fast UV-positive zone contains oxalate. This assumption is strengthened by the following experiment. When $FeCl_3$ was added to the oxalate
solution in increasing amounts, the width of the UV-positive zone
increased progressively (Fig. 2b-c)[+]. In aqueous solutions of oxalate
to which ferric ions have been added, the addition of complexing agents
(EDTA or EGTA, 10 mM) dramatically reduced the fast UV-positive zone
(Fig. 2d). However, addition of EDTA to aqueous solutions[++] of oxalate

---

[+] This has been independently shown by Frederiksson (ref. 12)
[++] Without added Fe^{3+}

only slightly reduced the first UV-positive zone, indicating that complexes of a ferri-oxalate are only in part responsible for this first fast-migrating zone.

Based on the assumption that both the UV-positive and the UV-negative zones contain oxalate in approximately identical concentrations. The combined zone length was measured at various oxalate concentrations. Linearity between oxalate concentration and combined zone length was observed over a range of concentrations from 0.2 to 10 mM oxalate in aqueous solutions (r = 0.998).

Reproducibility of measurement was checked by carrying out ten repeat determinations with different samples containing oxalate (injected volume 2 μℓ, oxalate content 1 nM). A coefficient of variation (C.V.) of 1.2 % was found. When repeat determinations were carried out under identical conditions with urine, to which oxalate had been added, a C.V. of 2.8 % was obtained. We consider that reproducibility can still be considerably improved with the use of automatic pipettes and larger injection volumes.

The specificity of the method was investigated by adding a number of organic acids to oxalate-containing urine samples. None of the acids (Table 1) interfered with the isotachophoretic measurement of oxalate. This is particularly noteworthy since some of these acids have been shown to interfere with colorimetric measurements of oxalate or measurement of oxalate by gas chromatography.

As expected, K-tetraoxalate interfered with oxalate determination, but this oxalate polymer does not exist in urine. Oxalate condensates, such as parabanic acid and oxaluric acid, did not interfere with the isotachophoretic measurement of oxalate. These acids may interfere with methods of oxalate determination that use pretreatment steps because of hydrolysis of the oxalate condensates.

Separation of oxalate from other organic anions is such that other organic anions may be determined together with oxalate in the same isotachophoretic run. This is demonstrated in Fig. 3, where the simultaneous determination of orotic acid and oxalic acid is shown.

TABLE 1

Interference in the isotachophoretic detection of urinary oxalate
by other acids

_____

No interference

Phthalic   Glucuronic
Tartronic   Saccharic
Glyoxylic   Mesaconic
Glycolic   Malonic
Orotic   Lactic
Tryptophan   Fumaric
Hippuric   Citric
Succinic

Possible interference:

Parabanic (= oxalil urea)
Oxaluric (= monoureide oxalic)

Definite interference:

K-tetra-oxalate

_____

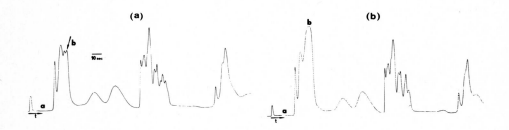

Fig. 3. (a) Complete isotachophoretic recording of untreated urine
            ($ZnCl_2$ 10 mM) with a high oxalate concentration
            (727 μmoles/l).
        (b) Same urine after addition of orotic acid (0.2 mM/l).
            a: UV-negative zone of oxalic acid; b: orotic acid.
            Injected volume, 6 μl; operational system as above; running-
            time 40 min.

Determination of oxalate in urine

    As in aqueous solutions, there was a UV-positive and a UV-negative
zone containing oxalate - both disappeared on treatment with oxalate

decarboxylase. If ferric ions are added to urine the fast migrating first UV-positive zone is lengthened, an effect which again can be completely reversed by the addition of EDTA. However, on the addition of EDTA to urine to which no ferric ions have previously been added, only a modest reduction of the first UV-positive zone can be obtained. It is concluded from these experiments that other substances released from glass or steel surfaces of the syringe interfere with oxalate to form soluble, fast-migrating oxalate-containing complexes which cannot be influenced by EDTA. However, this zone clearly contains oxalate, since treatment of urine samples with decarboxylase abolishes both the leading UV-positive and the UV-negative zones. This assumption is further supported by the observation that the width of the UV-positive zone increased considerably with syringes that had been in use for a long period of time compared to syringes not used before; or compared (qualitatively) to plastic syringes. Using new Hamilton syringes, linearity between oxalate concentration and the combined width of the leading UV-positive and the subsequent UV-negative zones was observed as shown in Fig. 4. This graph demonstrates that isotachophoresis can be used successfully to measure oxalate in urine. Complete recovery of oxalate only was obtained when $ZnCl_2$ was added to urine. Zinc may form complexes with oxalate, from which the isotachophoretic migration process can be easily started. Without added zinc, calcium oxalate crystals may be present, which may not be completely dissolved during the isotachophoretic run. Recovery of oxalate in urine samples was determined by adding increasing amounts of oxalate to 17 urine samples up to a final concentration of 1 mM added oxalate. All measurements were carried out at a constant injection volume (2 µℓ). The results (Table 2) clearly show linearity between added and recovered oxalate under these conditions. These values compare favourably with data reported in the literature (Table 3). However, it is clearly shown that the injection volume must be increased for urine samples containing low amounts of oxalic acid.

TABLE 2

Recovery of oxalate in urine samples (n = 17) at constant injection
volume (2 μℓ)

Oxalate added (nmoles/μℓ)	Recovered (mean %)	C.V. (%)	Range (%)
0.2	89.6	25.4	54-123
0.5	97.7	13	73-112
0.8	101.4	8	85-111
1.0	102	6.1	90-111

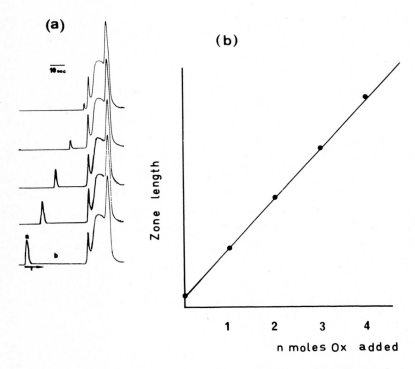

Fig. 4. To fresh untreated urine (ZnCl$_2$ 10 mM) oxalate was added in
increasing amounts (injected amounts of added oxalate: 0-1-2-
3-4 nmoles)
Linearity between the combined zone lengths of oxalate (a + b)
and the amount of added oxalate was observed.

TABLE 3

Recently published methods for determination of urinary oxalate

Method	Reference	Recovery of added oxalate (%)
Decarboxylase (EC 4.1.1.2.) + NAD	5	99.6 ± 0.44 (S.E.) 99.6 ± 0.44 (S.E.)
Decarboxylase + ^{14}C-Oxalate	6	98.4 ± 2 (S.E.)
Decarboxylase, change in pH	7	not given
Colorimetric, after co-precipitation with ^{14}C-oxalate	8	100.3 ± 2.5 (S.D.)
Colorimetric, after precipitation with calcium sulphate	9	85.6 ± 7.2 (S.D.)
Colorimetric, after anion-exchange column	10	98 ± 2 (S.D.)

CONCLUSIONS

These preliminary results demonstrate that isotachophoretic determination of urinary oxalate is possible without pretreatment of urine samples. For routine use in the clinical laboratory, 1 nM oxalate (i.e. 4-10 µl urine) should be injected. An even higher sensitivity may be achieved for reserach purposes when long capillaries or counterflow conditions are used. To avoid the formation of oxalate complexes it would be advantageous to use non-deformable plastic syringes ot syringes with Teflon surfaces and platinium needles.

ACKNOWLEDGEMENT

This study was supported by the Deutsche Forschungsgesellschaft, Ts 15/3.

REFERENCES

1 H.H. Poliers and P. Levatin, J. Biol. Chem. 154(1944)207.
2 P.M. Zarembski and A. Hodgkinson, Biochem. J., 96(1965)218.
3 H.E. Archer, A.E. Dormer, E.F. Scowen and R.W. Watts, Clin.Science, 16(1975)405.
4 R.A. Chalmers and R.U.E. Watts, Analyst, 97(1974)958.
5 J. Costello, M. Hatch and E. Bourke, J. Lab. Clin. Med., 87(1976) 903.
6 D.J. Bennet, F.E. Cole and E.D. Frohlich, J. Lab. Clin. Med., 91 (1978)822.
7 P.C. Hallson and G.A. Rose, Clin. Chim. Acta, 55(1974)29.

126

8 H. Baadenhuijsen and A.P. Jansen, Clin. Chim. Acta, 62(1975)315.
9 A. Hodgkinson and A. Williams, Clin. Chim. Acta, 36(1972)127.
10 F.M.F.G. Olthuis, A.M.G. Markslag, J.T. Klein Elhors and P.G.L.C.
   Krugers Dagneaux, Clin. Chem. Acta, 75(1977)123.
11 J.L. Beckers and F.M. Everaerts, J. Chromatogr., 69(1972)165.
12 S. Fredriksson, J. Chromatogr., submitted to publication

A. Adam and C. Schots (Editors), *Biochemical and Biological Applications of Isotachophoresis*   127
© 1980 Elsevier Scientific Publishing Company, Amsterdam — Printed in The Netherlands

ISOTACHOPHORESIS OF UREMIC METABOLITES

Frans Mikkers

Department of Instrumental Analysis,

Eindhoven University of Technology,

Eindhoven (The Netherlands)

and

Severin Ringoir

Department of Medicine, Nephrological Division,

University Hospital, University of Ghent,

B 9000 Gent (Belgium)

SUMMARY

   Isotachophoresis was used for the analysis of ionic metabolites in
the body fluids of patients with chronic renal failure.
Separation profiles of the sera of uremic patients are distinctly
different from the profiles of normal sera. From the profiles it is
seen that uremic patients form a heterogeneous group.
Using different operational systems it is shown that the major differences occur in the relatively low-molecular-weight anionic solutes.
Uremic urines and ultrafiltrates give similar separation profiles,
although relative concentration differences exist.
Examples of anionic separations in four electrolyte systems are given.

INTRODUCTION

   Patients with chronic renal failure generally have to be submitted
to intermittent hemodialysis or other dialysis and ultrafiltration
procedures. The patients show a complex of symptoms, usually called
"the uremic syndrome" or "uremia". Many of these symptoms are related
to a disturbance in the homeostatic or regenerative function of the
kidneys, which results in retention of metabolic products and in disorders of hormonal and metabolic function. By removing, even partially,
the accumulating waste products, the clinical manifestations of uremia
can be alleviated and many of the patients can pursue life in an "acceptable" way. Adequate dialysis implies the removal, or even retention,
of the right substances in the right way and to the right extent.

Though this may seem to be a problem, of medical technology, it is obvious that analytical chemistry can provide important "diagnostic" tools. With the increasing refinement in analytical chemistry the number of chemical substances that have been shown to accumulate during chronic renal failure has grown vastly. Nevertheless, it has not been possible to attribute the broad spectrum of uremic symptoms to the accumulation of known chemical substances (ref. 1). This could be due to the fact that the criteria for a uremic toxin are not uniform, whereas in addition, additive, antagonistic and synergetic effects may play an important role. Lack of knowledge of the nature of the substances that should be removed or retained hampers the development of new concepts of dialysis and of dialysis technology. It is therefore important that analytical methods are developed or applied in order to obtain reliable information for a number of biochemical parameters. The results of such research may have spin-offs in several ways. They should lessen our ignorance (ref. 2) of which substances are important and should increase our knowledge of metabolism and homeostasis. Analytical data can be directly used in physiological transport models in which a number of questions about efficiency can, at least qualitatively, be answered and through which simulation of dialysis procedures becomes possible. It is obvious that "new" developments in dialysis technology can be evaluated using reliable analytical parameters. Moreover, they provide the basic requirements for a more individual approach to the uremic patient.

The search for uremic toxins has been highly specific, the possible toxicity of single known substances being assessed. Due to the complex and non-specific nature of uremic metabolites and their possible interrelations such studies are rather elaborate and difficult. In recent years, however, some research groups have followed a screening approach to the problem, in which chemical multicomponent analysis of biological fluids derived from uremic patients plays an important role (refs. 3-7). The first step in this approach is to profile the body fluids of uremic patients into "fingerprints" of sufficient detail. Hence the analytical methods should be able to provide reliable information on a vast range of substances, with moderate differentiation. Using medical and biochemical techniques differentiation can be attempted. Analytical chemistry techniques have been developed that permit separation, identification, quantification and even isolation of many metabolites in biological fluids. The choice of which technique to apply is in practice only limited by financial and man-power facilities. Due to the large variety of uremic solutes several analytical techniques have to be used. In our screening approach four different techniques are used : (1) gas chromatography, (2) mass spectrometry, (3) liquid chromatography, and (4) electrophoresis. For the identification of solutes and to increase the differentiating capabilities the various techniques can be combined. Serum, urine and ultrafiltrate samples are submitted to profiling, though other body fluids such as dialysate, etc., are not excluded. Many choices of methodology have to be made since the accumulating solutes can be very different : volatile non-volatile, low molecular weight high molecular weight, ionic non-ionic. For the qualitative and quantitative evaluation of ionic solutes several analytical techniques are available, each with its own disadvantages and limitations. Isotachophoresis is compatible with the basic requirements for screening procedures, namely, multicomponent information, rapid completion, reliability and low cost. Moreover, the flexibility of the technique offers a wide range of possibilities for detailed studies. Hence, isotachophoresis shloud provide useful information on the occurence of ionic solutes in uremic biological fluids.

MATERIALS AND METHODS

All chemicals used were of analytical grade or purified by conventional methods. Isotachophoresis was performed using the equipment developed by Everaerts et al. (ref. 8). The separation compartment consisted of polytetrafluoroethylene narrow-bore tubing with an inner diameter of 0.2 mm and an outer diameter of 0.4 mm. The direct, constant electrical driving current was taken from a Brandenburg (Thornton, Healh, Great Britain) high-voltage power supply. Separated zones were detected by measuring the electrical conductivity as well as the UV absorption at 254 mm. Both detection systems can be used for quantitative and qualitative evaluation of the zone characteristics. All samples were analysed for their anionogenic contents.

Electrolyte systems and other operational conditions are summarized in Tables I, II and III. To increase the experimental performance 0.2 % hydroxyethylcellulose (PolySciences, Warrington, Pa., U.S.A.) was used as a viscous additive. Separation times were less than 20 min. In contrast with many other separation techniques isotachophoresis generally needs no sample pre-treatment and samples can be directly introduced by means of a microliter syringe.

Serum, urine and ultraflitrate samples were taken from a 36-year-old
female patient with chronic glomerulonephritis, on intermittent hemodialysis since March 1975. The glomerular filtration rate was less
than 2 ml.
From January 1978 the patient was treated with a sequential 1-h ultrafiltration, 3-h diffusion procedure following a modified Bergström
technique (ref. 9).
To increase the blood compartment pressure and the transmembrane pressure in the 1 m^2 PAN dialyser (RP6), a double-head pump (Bellco) was
used. Reference samples were taken from a healthy female subject. The
reference ultrafiltrate was obtained by pressure ultrafiltration of
the normal serum through an Amicon PM 10 membrane (molecular weight
cut-off 10,000).

RESULTS AND DISCUSSION

In using isotachophoresis as a screening procedure much information
must be obtained in a relatively short analysis time. We therefore
limited the dialysis time to 20 min at moderate current densities.
As regards the multicomponent information a methodological choice has
to be made between anionic and cationic separations. Preliminary experiments showed that cationic profiling of uremic sera gave only
minor differences; creatinine, for example. On the other hand, anionic
separations showed more deviating results. So we confined our investigations to the anionic separation mode.

Anionic samples solutes will have sufficient electrophoretic mobility, provided the pH of the leading electrolyte is high enough. For
the anionic profiling of uremic sera (Fig. 1) a chloride-hydroxyl
system was used. The operational conditions for this electrolyte system
are given in Table I. The performance characteristics of this system
are better than those of conventional electrolyte systems, in which
an amino acid of doubtful purity is generally used as the terminator.
Fig. 1 shows the analytical results when 0.2 µℓ of serum, (normal,
uremic before and uremic after hemodialysis) is injected. Due to the
swamping amount of chloride and the relatively high protein content
only a low sample load can be applied. The differences between the
uremic and the normal serum are evident. The protein region, in which
albumin is the most abundant solute, shows only minor differences.
These can be due to the proteins or to low concentrations of some
amino acids, peptides, purines and pyrimidines, with fairly high pK_a

TABLE I

Operational conditions for anionic separations at high pH

Leading ion	Chloride
Concentration	0.01 M
Counter ion	$TRIS^+$ (0.01 M), NH_4^+
pH of leading electrolyte	8.10
Terminator	$Ba(OH)_2$/pH 11.5^{++}
Driving current	30.10^{-6}A, j = 0.0955 A/cm^2
Temperature	Ambient
Detection	Conductimetric
	UV 254 nm, transmission
Recorder	6 cm/min

+ TRIS = trishydroxymethylaminomethane; TRIS-HCl (Sigma) 0.01 M to
  pH 8.10 with ammonia.
++ OH is used as the terminating ion, enforced system.

values. The major differences, however, occur in the lower molecular
weight region and many of the accumulating solutes have UV absorption
at 254 nm. The somewhat poor differentiation in this region is the
analytical price for the successful attempt to stack as many anionic
solutes as possible. The large zone between lactate and phosphate is
mainly carbonate, originating from the electrolyte system and the
sample. On comparing the analyses before and after hemodialysis, it
is seen that the accumulating solutes have only been partially removed.
Moreover, acetate, originating from the dialysate, is clearly visible.

Better differentiated results are obtained when the pH of the leading
electrolyte is lowered. The electrolyte system at pH 6, which covers a
large spectrum of anionic solutes, is given in Table II. It should be
emphasized that many of the proteins, peptides, amino acids, purines
and pyrimidines will not migrate isotachophoretically in this system.
In general substances with a pK_a value higher than that of the terminator, pK_a (HEPES) = 7.55, will not be stacked. Since some solutes
are immobilized at this pH, the load capacity of the equipment will
increase. Since uremic sera can show appreciable fluctuations it is
wise to work well below maximum load capacity. Fig. 2 shows the separation porfiles when the same amounts (0.3 µl) of the different sera
are injected. Obviously the normal serum (Fig. 2A) is significantly
different from the uremic serum before the sequential procedure (Fig. 2B)
Differences between the uremic serum before (Fig. 2B) and after the sequential procedure (Fig. 2C) are much less pronounced. The anionogenic
contents of the normal ultrafiltrate and the normal urine are shown in

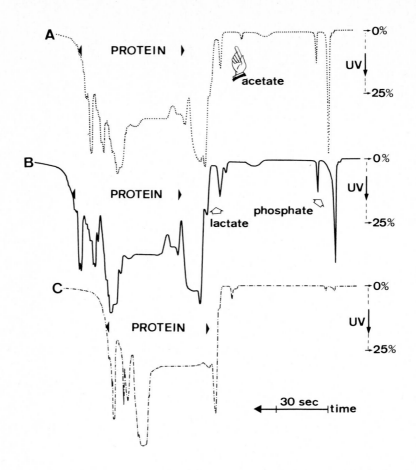

Fig. 1. Anionic profiling at pH 8.10. Operational system: Table 1.
A: uremic serum after dialysis, 0.2 µℓ. B: uremic serum before
dialysis, 0.2 µℓ. C: normal serum, 0.2 µℓ.

Fig. 3A and B, respectively. The ultrafiltrate and the urine of the
patient are given in Fig. 3C and D. The sample volume for each ultrafiltrate was again 0.3 µℓ, whereas for the urines 0.1 µℓ was injected.
From Figs. 2A and 3A it follows that through the ultrafilfration
procedure a substantial amount of solutes have been removed from the
normal serum. The ultrafiltrate of Fig. 3A contains only substances
with a molecular weight smaller than 10,000. These substances migrate
as rather mobile substances.
The same solutes are present in the normal serum, indicated in Fig. 2
by the L-region. The higher-molecular-weight fraction, H-region, is
not present in the ultrafiltrates and urines. Zone N° 22 represents

TABLE II

Operational conditions for anionic separations at pH 6

Leading ion	Chloride
Concentration	0.01 M
Counter ion	Histidine
pH of leading electrolyte	6.02
Terminator	HEPES$^+$
Driving current	20.10^{-6} A, $j = 0.0637$ A/cm^2
Temperature	Ambient
Detection	Conductimetric
	UV, 254 nm, transmission
Recorder	6 cm/min

+ HEPES = N-2 hydroxyethylpiperazine-N'-2-ethane sulphonic acid
  (Serva), 0.02M with TRIS to pH 8.

the rejection point of the ultrafiltration membrane. The most important
solute of the higher-molecular-weight region is albumin, though other
substances of high molecular weight are presumably present. Comparing
normal urine and ultrafiltrate it follows that they both comprise the
same mobility spectrum. As a result of the unique selective concentrating capabilities of the normally functioning kidney the analyses
of the normal urine and the ultrafiltrate show distinct, different
separation profiles.

From the electropherograms in Fig. 2 the differences can be seen
between a normal serum and an uremic serum. Clearly there is an excess
of rather mobile solutes in the L-region. These substances include
sulphate, indoxylsulphate, adenosine triphosphate, orotate, hippurate,
glucuronate and urate. For some of these solutes concentration differ
by an order of magnitude from normal values. Much less-pronounced
differences are present in the H-region, which seems to be somewhat
more heterogeneous and apparently contains more mobile solutes. After
the dialysis procedure (Fig. 2C) the acetate (zone N° 6) originating
from the dialysate is clearly visible. From the decrease in the zone
lenghts and the decrease in the UV absorption it can be concluded
that several solutes have been removed to some extent. Nevertheless
the electropherogram still differs significantly from the normal one.
A change has also occured in the higher molecular weight region, which
seems to be more homogeneous. Moreover, the albumin fraction shows
an increased electrophoretic mobility. Although the exact reason for
this behaviour is not clear, it is specific for the post-treatment
state.

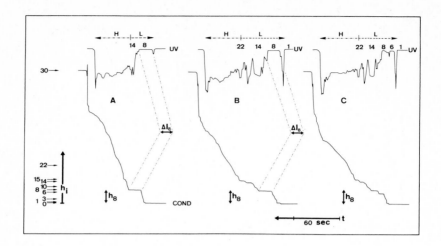

Fig. 2. Isotachophoretic serum profiles at pH 6.
Operational system: Table II. UV = UV (254 nm) absorption,
COND = conductivity, t = time, H = higher-molecular-weight
region, L = lower-molecular-weight region. A: normal serum,
0.3 µℓ. B: uremic serum before sequential procedure, 0.3 µℓ.
C: uremic serum after sequential procedure, 0.3 µℓ. Zones =
0 = chloride, 1 = sulphate, 3 = indoxysulphate + adenosine
triphosphate, 6 = acetate, 8 = phosphate + lactate , 10 =
hydroxybutyrate, 14 = hippurate, 15 = glucuronate, 30 = HEPES
(terminating ion).

Fig. 3C shows the ultrafiltrate at the beginning of the sequential
procedure. Due to the fact that dialysate was used for the initial
wetting of the membrane, some acetate (zone N° 6) is present. Obviously
the ultrafiltrate contains the same low-molecular-weight solutes as
the serum. Minor concentration differences are due to membrane polarization. Whereas normal serum and urine show distinct, different
separation profiles, patients' urine and ultrafiltrate show a marked
resemblance. The concentration differences are partly due to the renal
rest function. Especially hippurate, zone N° 14, is present in a
relatively high concentration. Although the uremic patient's urine
gives a separation profile that is different from the normal one, this
difference is not significant since large fluctuations in normal urines
occur. Fig. 3 shows in addition the ultrafiltrate of normal serum
(Amicon, PM 10, mol.wt. cut-off 10,000). It is obvious that lactate
and phosphate aee the most abundant anions, other than chloride, in
normal serum.

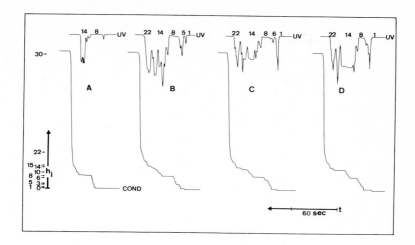

Fig. 3. Isotachophoretic ultrafiltrate and urine profiles at pH 6.
A: normal ultrafiltrate, 0.3 µℓ. B: normal urine, 0.1 µℓ.
C: uremic ultrafiltrate, 0.3 µℓ. D: uremic urine, 0.1 µℓ.

Since the excess of lower-molecular-weight substances is a characteristic difference between the normal and the uremic state, it is
advantageous to describe this difference as a single quantitative
parameter. The ratio of anionic higher-molecular-weight substances
to anionic lower-molecular-weight substances, the HL ratio, seems
appropriate, the ratio being given by the quotient of the two subtotal
zone lengths in the UV profile. It should be emphasized that the ratio
has no absolute value and is system-dependent. Moreover it should be
mentioned that even high-melocular-weight substances such as heparin
can migrate as rather mobile substances, provided they have a large
net electrical charge. As a criterion for finding the ratio we chose
zone 22, since this zone is characteristic of all separations done until
now and forms the rejection point of a PM 10 membrane. On screening
a patient for several months, it was found that the ratio before and
after hemodialysis was fairly constant (ref. 5).

As can be expected, patients with chronic renal failure do not form
a homogeneous group. Fig. 4 gives some separation profiles of different
patients before hemodialysis. Patients with a very large excess of
anionic lower-molecular-weight solutes will have a low ratio, HL = 0.68.

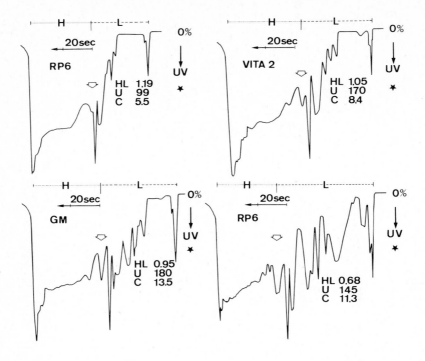

Fig. 4. Anionic separation profiles of several patients.
Operational system: Table II. Sample volume: 0.3 µℓ. H: higher-
molecular-weight region. L: lower-molecular-weight region.
HL: HL-ratio. A: urea (mg/100 ml). C: creatinine (mg/100 ml).
RPG, VITA 2 and GM: dialysers.

As can be seen, the separation profiles can be very different. The
example with HL = 1.19 hardly deviates from normal. The samples also
show differences in the lower-molecular-weight part, L, as in the
higher-molecular-weight part, H. The differences in the latter part,
however, are much less pronounced. In fig. 4 the urea (mg/100 ml) and
creatinine (mg/100 ml) concentrations and the makes of dialyser are
also given. From the date it follows that the high HL ratio is in good
agreement with the fairly low urea and creatinine concentrations.
HL ratios generally improve after hemodialysis and can even exceed
the normal value.

Whereas the electrolyte system at pH 6 can be used for a fast screening approach, more differentiation is obtained at low pH. The operatio-

nal system for anionic separations at pH 3.5 is given in Table III.

TABLE III

Operational conditions for anionic separations at pH 3.5

Leading ion	Chloride
Concentration	0.01 M
Counter ion	β-alanine
pH of leading electrolyte	3.50
Terminator	Caproate
Driving current	25.10^{-6} A, j = 0.0796 A/cm^2
Temperature	Ambient air
Detection	Conductimetric
	UV, 254 nm, transmission
Recorder	6 cm/min

Fig. 5 gives the separation profiles of a normal and a predialysis sample. Whereas in the normal serum separation lactate and phosphate are the most abundant solutes, in the uremic sample solutes with UV absorption are present in addition. Caproate was used as the terminating ion and therefore in general only substances with a pK value lower than 4.83 will migrate isotachophoretically.

Since many of the solutes are now immobilized, the sample load could, in comparison with pH 6, be further increased. At this pH of the leading electrolyte there are no significant differences between ultrafiltrates (PM 10) and sera. We investigated the influence of the ultrafiltration membranes Amicon XM 50, PM 30, UM 20, PM 10, UM 10, YM 10, UM 2 and UM 05. Down to PM 10, mol.wt. cut-off 10,000 no significant differences were measured. However, using membranes with an even lower rejection point selectivity effects were observed. Relative concentrations of very-low-molecular-weight substances such as phosphate, sulphate, and lactate did change. Hence it should be concluded that polarization effects at the membrane surface can play an important role. Fig. 6 gives the separation proifles of serum of the same patient after hemodialysis. In comparison with the pre-dialysis sample the excess of solute, as analysed by isotachophoresis, has been reduced to ± 60 %. Acetate from the dialysate is clearly visible.

Very selective determination can be made using the UV spike method. Fig.7 gives an example for the determination of uric acid. The operational system was chosen in such a way that most of the sample solutes migrated zone electrophoretically either in front of or behind the isotachophoretic stack. p-Aminobenzoic acid was used as the internal

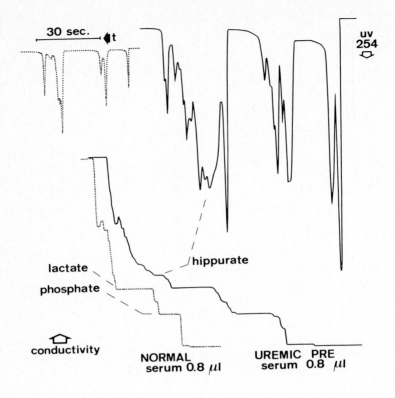

Fig. 5. Anionic profiling at pH 3.50

standard and valproate was used to space the two UV-absorbing solutes
isotachophoretically. The UV signal, measured at 280 nm, was electronically converted to extinction and the peak areas were integrated
by a fast sampling electronic integrator. Linear calibration lines
were obtained and the detection limit is below 1 μ mole. By the use
of the internal standard method, injection faults are excluded.
Although the developemnt of the appropriate electrolyte systems is
not always easy the method seems promising for the fast determination
of low concentrations of UV-absorbing solutes.

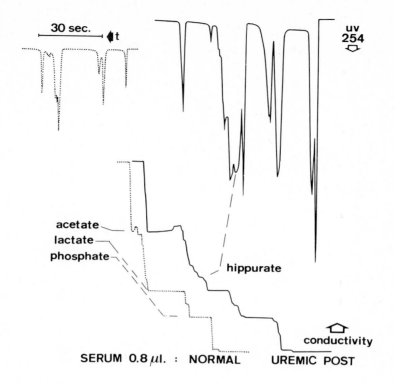

Fig. 6. Anionic profiling at pH 3.50

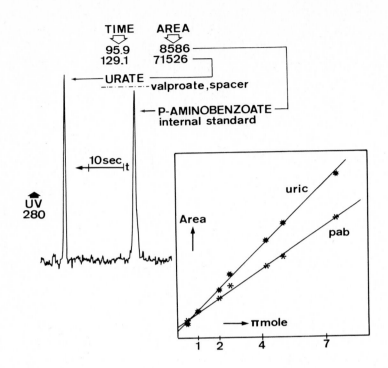

Fig. 7. Selective determination of urate.
Leading ion: adipate,0.0025 M. Counter ion: TRIS.
pH of the leading electrolyte: 4.8. UV: extinction at 280 nm.

REFERENCES

1 J. Bergström and P. Fürst, Kidney Int., 15(1978)9.
2 C. Schreiner, Kidney Int., 7(1975)270.
3 J. Bergström and P. Fürst, Clin. Nephrol., 5(1976)143.
4 F. Senftlebre, A. Halline and H. Veening, Clin. Chem., 22(1976)1522.
5 F. Mikkers, S. Ringoir and R. de Smet, J. Chromatogr., 162(1979)341.
6 F. Bultitude and S. Newham, Clin. Chem., 21(1975)1329.
7 A. Schoots, F. Mikkers, C. Cramers and S. Ringoir, J. Chromatogr.,
    in press.
8 F. Everaerts, J. Beckers and Th. Verheggen, Isotachophoresis,
    Elsevier, Amsterdam, Oxford, New York, 1976.
9 J. Bergström, H. Asaba, P. Fürst and R. Oulès, Proc. EDTA, 13(1976)
  293.

A. Adam and C. Schots (Editors), *Biochemical and Biological Applications of Isotachophoresis*   141
© 1980 Elsevier Scientific Publishing Company, Amsterdam — Printed in The Netherlands

ISOTACHOPHORETIC SEPARATION OF MIDDLE MOLECULE PEPTIDES IN UREMIC BODY
FLUIDS

L. Zimmerman, A. Baldesten, J. Bergström and P. Fürst[+]
Metabolic Research Laboratory and Department of Nephrology, S:t Erik's
Hospital, Stockholm (Sweden), and Application Laboratory, LKB-Produkter
AB, Bromma (Sweden)

INTRODUCTION AND BACKGROUND

The term uremia, originally coined by two Frenchmen, Piorry and
l'Heriter in 1840, literally means urine in the blood.

According to the chemical theory of uremic toxicity, it is assumed
that substances that are normally excreted by the kidneys are retained
in the body fluids and exert toxic effects. Toxic manifestations of
uremia seem to affect practically all organs and tissues of the body;
mucous and serous membranes are influenced, various transport phenomena
in the cell membranes are inhibited, and intermediary metabolic processes are impaired.

It has long been known that uremic toxicity is a reversible phenomenon. Certain therapeutic regimes may decrease or abolish symptoms
of uremic toxicity. Maintenance dialysis has for many years been a
well-established method of treating renal failure. It is remarkable
that we still know so little of what the artificial kidney really
does in terms of removing toxic metabolites. Dialysis as performed
today is a relatively unselective procedure. It cannot, however, be
made more selective until we know exactly that we want to get rid of.

Lacking more precise knowledge of which compounds are critically
toxic in uremia, pragmatism has been the guideline for progress in
research. One critical factor in any system which is based on diffusion
or filtration through semipermeable membranes is molecular size.
Accordingly it has now become the fashion to classify the uremic solutes
on the basis of their molecular weights.

[+] Mailing address: Metabolic Research Laboratory, S:t Erik's Hospital
    S-112 82 Stockholm, Sweden.

The middle molecule hypothesis was originally developed by Schribner and co-workers (ref. 1 and 2) to explain the clinical observation that patients treated with intermittent peritoneal dialysis did not develop neuropathy as readily as hemodialysis patients in spite of being less efficiently dialyzed with regard to small molecules such as urea and creatinine. It was assumed that substances with a molecular weights between 500 and 5000, so called "middle-molecules", were toxic and that the peritoneal membrane was more permeable to these middle molecules than the membranes in the artificial kidney (ref. 3). Few new theories in nephrology have evoked so much interest and created so much confusion as the middle molecule hypothesis.

In order to answer all the questions raised related to this hypothesis we need reliable methods for the estimation and identification of the middle molecules. In the last five years several groups have been able to demonstrate that ultraviolet-absorbing solutes in the middle molecule range accumulate in the plasma of patients with severe uremia (ref. 4). However, with the gel-filtration methods commonly used it was only possible to obtain a crude separation of middle molecule compounds, which yielded at best semi-quantitative results (ref. 5).

Our research group has for many years been engaged in research related to uremic middle molecules. High-speed-gel-filtration chromatography (HSGF) on Sephadex G-15 columns enabled us to separate normal and uremic plasma, urine, and dialysis fluid into 7-8 UV-absorbing peaks. One of these peaks (n° 7) corresponded to a molecular weight range of 1000-2000. This peak is present in the plasma of patients with severe uremia but is not detected in normal plasma. The peak 7 material isolated by the gel-filtration procedure could be further separated into 7-9 subfractions (peaks 7a, b, c, etc.) by gradient-elution ion-exchange chromatography (GEC) on columns with DEAE-Sephadex A-25 eluted with a TRIS or ammonium acetate buffer. These methods are described in detail in a previous communication (ref. 6). Most of these subpeaks were present in plasma from patients with severe uremia, but only two of them, 7f and g, were detected in normal plasma (refs. 4 and 5).

In the search for a technique allowing further separation and purification, the isotachophoretic method (refs. 7 and 8) has been tried with the expectation that it would offer advantages over previously described techniques. In the present communication further separation of three of the uremic middle molecule subpeaks by isotachophoresis is described. Each of the zones separated corresponds to a single middle molecule compound.

MATERIAL AND METHODS

Filtered urine or ultrafiltrate from severely uremic patients on intermittent hemodialysis treatment was used. The ultrafiltrate was obtained at the beginning of a hemodialysis treatment by applying negative pressure in the dialysate compartment while no dialysis fluid was running through the dialyzer, using a dialysis monitor modified for sequential ultrafiltration dialysis (Gambro UDM) (ref. 9). Normal urines were obtained from laboratory personnel.

The analytical and preparative procedures applying the combined HSGF-GEC technique and using TRIS buffer have recently been described in full (ref. 6).

Analytical isotachophoresis

The isotachophoretic separations and determinations were made on an LKB 2127 Tachophor. The separations required about 30 min and the current at detection was 65 μA.

Electrolytes

The chemicals used were analytical grade. The water used was double glass distilled and ion exchanged.

Leading electrolyte : 5 mM HCl Surprapur (E. Merck, Darmstadt, G.F.R.), 20 mM β-alanine (Sigma, St. Louis, Mo., U.S.A.), and 0.4 % HPMC (Dow Chemical, Midland, Mich. U.S.A.) (pH 4.0).

Terminating electrolyte : 5 mM caproic acid (Fluka, Buchs, Switzerland).

RESULTS AND DISCUSSION

Isotachophoretic analyses were performed on the crude middle molecule fraction (peak 7) in normal and in uremic biological fluids. Middle molecule patterns of a normal and a uremic urine are recorded in Figs. 1 and 2. As a comparison the corresponding resolution using ion-exchange chromatography is recorded in the figures. The crude middle molecule fraction consisted of several UV-absorbing substances, as previously reported (refs. 5 and 6). Additionally, the isotachophoretic pattern indicated the presence of non-UV-absorbing solutes. The majority of the zones were too narrow to be detected and separated.

Peak 7 Normal Urine

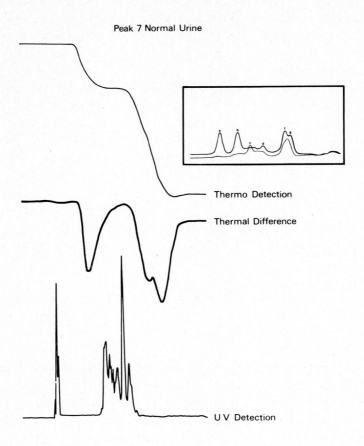

Thermo Detection

Thermal Difference

U V Detection

Fig. 1. Isotachophoretic separation of peak 7 material isolated by molecular size separation (HSGF) from normale urine. The detection was made at 65 µA, 254 nm and with a chart speed of 5 cm/min. 1 µl of peak 7 was injected, corresponding to a urine volume of 2 µl. As a comparison the corresponding resolution is recorded using gradient elution chromatography (GEC).

It was also obvious from the curves that certain compounds which were observed in uremic urine were not recovered in normal urine. The most distinct UV-absorbing zone in uremic urine, commonly very prominent was not found in normal urine.

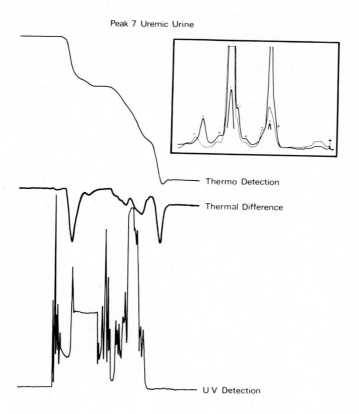

Peak 7 Uremic Urine

—— Thermo Detection

—— Thermal Difference

—— U V Detection

Fig. 2. Isotachophoretic separation of peak 7 material isolated by
molecular size separation from a uremic urine. The detection
was made at 65 μA, 254 nm and with a chart speed of 2 cm/min.
2 μl of peak 7 were injected, corresponding to a urine volume
of 2 μl. As a comparison the corresponding resolution is
recorded using gradient elution chromatography (GEC)·

As reported earlier, peak 7 material was further separated by the
combined technique into 7-8 UV-absorbing subpeaks (refs. 5 and 6).
Prior to isotachophoresis three of the respective subpeaks(7a, b,
and c) were recycled using combined gel filtration-ion-exchange chromatography to check the purity of the isolated peak material, as

illustrated in Fig. 3. Each subpeak was homogeneous and appeared at
the expected retention volume. The peaks detected were symmetric and
no sign of overlapping material was observed except a slight shoulder
on peak 7b. Thus, it was not possible to separate the subpeaks further
by gradient ion-exchange chromatography, nor was any improvement achieved by rerunning the "chromatographically homogeneous" fractions shown
in Fig. 3.

Isotachophoretic analysis of subpeak 7a resulted in two main zones
with UV levels of 97 % and 15 %[+], respectively (Fig. 4a).

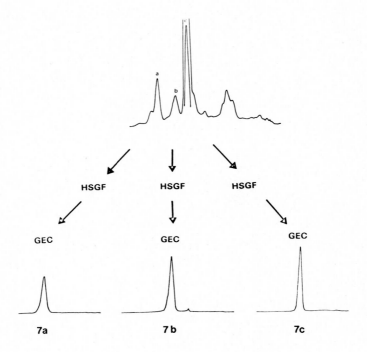

Fig. 3. An illustration of the procedure used to test the purity of
the subpeaks isolated by a preparative technique. The respective subpeaks (7a, b, and c) were recycled using the
combined analytical method.

_____

+ As a percentage of full scale.

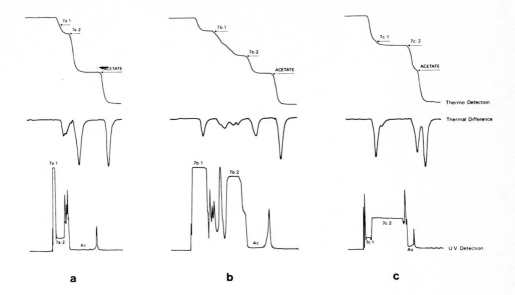

Fig. 4. Isotachophoretic separation of peaks 7a, b, and c, isolated
       from a uremic ultratiltrate by a preparative technique using
       acetate buffer. The thermal step heights are indicated in the
       figure as well as the corresponding UV levels. 10 µl of peak
       7a, 10 µl of peak 7b, and 2 µl, of peak 7c were injected,
       corresponding to 200 µl, 1600 µl, and 14 µl of ultrafiltrate
       respectively. The detection was made at 254 nm, 65 µA and with
       a chart speed of 2 cm/min.

The thermal step height for the first zone was 8 % and for the second
zone 19 %[++]. A non-UV-absorbing zone with a thermal step height of
64 % showed the presence of acetate, i.e. a residue from ammonium
acetate buffer.

   The isotachophoretic analysis of subpeak 7b showed that this subpeak
consists of several UV-absorbing solutes as recorede in Fig. 4b. The
two main zones had UV levels of 97 % and 86 % and thermal step heights
of 14 % and 43 %, respectively. Between the main zones minor UV-absorbing peaks indicated the presence of other solutes. A non-UV-absorbing

++ As a percentage of the mobility of the terminator

zone from acetate was present.

Isotachophoretic analysis of subpeak 7c revealed two zones close
to each other in net mobilities   (Fig. 4c). They have UV levels of
16 % and 38 % and thermal step heights of 30 % and 34 %, respectively,
The characteristic values for middle molecule fractions are summarized
in Table 1.

TABLE 1

Characteristic values for middle molecule fractions

	Mobility[+]	UV level[++]
7c:1	30 %	16 %
7c:2	34 %	38 %
7b:1	14 %	97 %
7b:2	43 %	86 %
7a:1	8 %	97 %
7a:2	19 %	15 %

+ As a percentage of the mobility of the terminator.
++ As a percentage of full scale.

When evaluating single zones with different concentrations, excellent linearity was found, allowing the quantity of material to be
calculated by measuring the zone length. The zone lengths detected
and measured for peak 7c material are recorded in Fig. 5. It was
possible to apply 1-10 μl, in which sample range the UV-absorbing zones
were detectable and directly proportional to the amount of material
injected. With larger amounts, equilibrium was not achieved with the
length of capillary used. In this case a new mixed zone occured between zones $7c_1$ and $7c_2$.

It is important to point out that identical zones as regards both
mobility and UV level were obtained when processing the isolated subpeak material from different uremic biological fluids, from different
patients and by applying different preparative procedures. Since in
isotachophoresis the concentration in each zone remains in equilibrium
with the leading ion, a constant thermal step height and UV level
must be obtained for a given substance on each occasion analysed.
During this investigation the reproducibility of parameters was better
than ± 2 % for the material studied. Consequently the material recovered
from different sources must be identical.

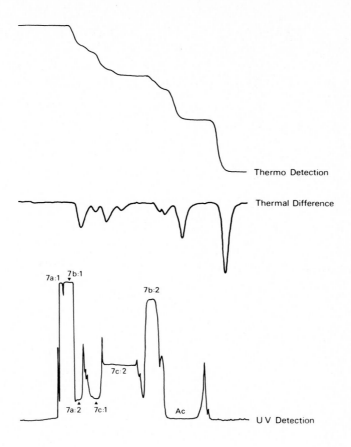

Fig. 5. Isotachophoretic separation of a mixture of peaks 7a, b and c.
The respective fractions were isolated in acetate buffer.
Analytical conditions are given in Fig. 4.

It can be concluded from the characteristic values summarized in
Table 1 that all six main zones corresponded to different substances.
This has been verified by dopping experiments, where all the six main
zones were mixed, analysed and recovered according to their respective
mobilities, as demonstrated in Fig. 6.

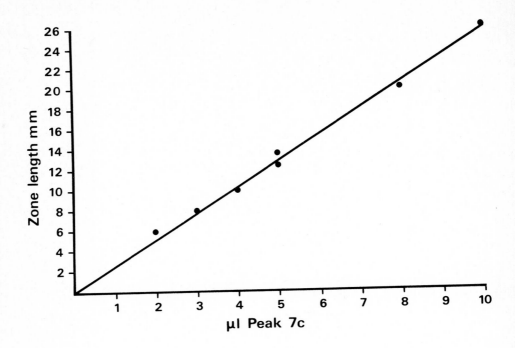

Fig. 6. The relation between zone length measured and volume of peak
7c injected.

All the separated solutes have a molecular weight in the range
500-2000 and represent substances with negative net charges. The
mobilities of the separated solutes in the electrolyte system used,
indicate pI values well below 4. From the position of the subpeak on
weak DEAE ion-exchange chromatography it can be concluded that the
levels of dissociation are $Diss_a > Diss_b > Diss_c$. From the isotachophoretic
runs the mobilities at pH 3.9 are in the orde $m_{a1} > m_{b1} > m_{a2} > m_{v1} > m_{c2} > m_{b2}$,
with surprisingly large steps between some of the substances, and not
in the expected sequence ($m_{a1} > m_{a2} > m_{b1} > m_{b2} > m_{c1} > m_{c2}$). This can be explained if the dissociation curves do not run in parallel or if there are
pH-dependent steric alterations in the molecules.

The chemical nature of these middel molecules is still insufficient-
ly known, since none of them has yet been obtained in a pure enough form
to enable structural analysis. In an attempt to clarify the chemical
structure of peak 7c material amino acid analyses were performed before

and after acid hydrolysis, and trials were also made to determine the N- and C-terminals. The main peptide material consisted of a polyglycine chain with blocked N-terminal.((Jörnvall et al., unpublished observation).

In recent studies amino acid analyses were also carried out before and after acid hydrolysis of peak 7a and 7b preparations, showing that peaks 7a and 7b contain peptidic material. The amino acid pattern and the proportions of the amino acids differed considerably from those seen for peak 7c.

Further characterisation of these materials requires suitable preparative technology. One of the possibilities may be the use of preparative isotachophoresis.

ACKNOWLEDGMENT

This work was conducted with support from NIH contract N° NO1-AM-2-2215.

REFERENCES

1 B.H. Schribner, Trans. Amer. Soc. Artif. Intern. Organs., 11(1965)29.
2 A.L. Babb, R.P. Popovich, T.G. Christopher and B.H. Schribner, Trans. Amer. Soc. Artif. Intern. Organs., 17(1971)81.
3 A.L. Babb, P.C. Farrell, D.A. Uvelli and B.H. Schribner, Trans. Amer. Soc. Artif. Intern. Organs., 19(1973)320.
4 J. Bergström and P. Fürst, Uremic toxins. In: W. Drukker, F.M. Parsons and J.F. Maher (Editors), Replacement of Renal Function by Dialysis, Martinus Nijhoff Medical Division, The Hague, Boston, London, 1978, pp. 334-368.
5 J. Bergström and P. Fürst, Clin. Nephrol., 5(1976)143.
6 P. Fürst; L. Zimmerman and J. Bergström, Clin. Nephrol. 5(1976)178.
7 H. Haglund, Isotachophoresis - Sci. Tools, 17(1970)1.
8 A.J.P. Martin, and F.M. Everaerts, Proc. Roy. Soc. London, 316(1970) 493.
9 S. Shaldon, H. Asaba, K. Lindh, P. Fürst, L.-A. Larsson and J. Bergström, T.H. Frost (Editor), Technical Aspects of Renal Dialysis. Pitman Press, Bath, Great Britain, 1978, pp. 282-288.

A. Adam and C. Schots (Editors), *Biochemical and Biological Applications of Isotachophoresis*   153
© 1980 Elsevier Scientific Publishing Company, Amsterdam — Printed in The Netherlands

THE ASSAY OF ENZYMATIC PROCESSES INVOLVING UDP-GLUCURONATE AS SUBSTRATE BY ANALYTICAL CAPILLARY ISOTACHOPHORESIS[+]

C.J. Holloway

Institut für Klinische Biochemie, Medizinische Hochschule Hannover, Karl-Wiechert - Allee 9, D3000 Hannover 61 (G.F.R.)

SUMMARY

The nucleotide UDP-glucuronate is utilised in at least two enzyme systems of the microsomal fraction of liver homogenate. The first, UDP-glucuronyltransferase (EC 2.4.1.17) catalyses the transfer of the glucuronate residue from the nucleotide to a wide range of xenobiotic and endogenous toxins, and serves as a vital detoxification mechanism in the liver and to a certain extent in other organs. UDP-glucuronate also serves as substrate for a hydrolytic enzyme, which is believed to be a non-specific nucleotide pyrophosphatase (EC 3.6.1.9), the products being UMP and glucuronate 1-phosphate. This enzyme probably fulfills a regulatory rôle in the so-called glucuronic acid pathway.

There are a large number of different assay methods available for UDP-glucuronyltransferase, most of which exploit a specific property of the acceptor molecule employed, or a property of the glucuronidated product, and take no account of the loss of donor molecule UDP-glucuronate through pyrophosphatase cleavage. To date, the only accurate method reported for the analysis of the pyrophosphatase has been through classical thin-layer chromatography, which is tedious, and requires labelled substrates for a satisfactory quantitative assay.

[+] Enzymes: UDP-glucuronyltransferase (UDP-glucuronate:glucuronosyl-transferase) (EC.2.4.1.17); nucleotide pyrophosphatase (dinucleotide nucleotidohydrolase) (EC 3.6.1.9); alkaline phosphatase (nucleotide nucleotidohydrolase) (EC 3.1.3.1). Abbreviations: UTP = uridine 5'- triphosphate; UDP = uridine 5'-diphosphate; UMP = uridine 5'-monophosphate; UDPG - uridine 5'-diphosphoglucose; UDPGA = uridine 5'-diphosphoglucuronate; GA-1-P = glucuronate 1-phosphate; G-1-P = glucose 1-phosphate; GA = glucuronate; P_i = inorganic phosphate.

Nonetheless, in all but the purest UDP-glucuronyltransferase preparations, it is important to be able to assess the effect of both
enzymic activities, especially when the determination of substrate
concentration effects is involved.

We report here the application of analytical capillary isotachophoresis to a complete analysis of UDP-glucuronate hydrolysis through
nucleotide pyrophosphatase activity, and the further hydrolysis of the
primary products UMP and glucuronate 1-phosphate through a general
alkaline phosphatase activity. The method has further been extended
to the assay of UDP-glucuronyltransferase, using paracetamol, a drug
of current toxicological interest, as glucuronide acceptor. This
acceptor is metabolised considerably more slowly than other phenolic
compounds by the rabbit liver microsomal UDP-glucuronyltransferase,
so that the relatively rapid cleavage of the nucleotide by the general
pyrophosphatase activity becomes of prime importance. Under the
conditions described here it is possible to assess the effect of both
activities simultaneously.

The analysis is accurate with as little as 1 nmole of a particular
substrate or product in the reaction aliquot, but a usable result is
obtained with as little as 0.1 nmole of substance in many cases. It
is suggested that this method can be extended for the assay of other
multiple enzyme systems involving nucleotides as substrates.

INTRODUCTION

Conjugates of endogenous and exogenous toxins with hydrophilic
residues such as glucuronate, sulphate, or glutathione are an important
aspect of hepatic detoxification mechanisms. Such pathways provide
not only the nullification of xenobiotic properties of a drug, but
also render the molecule sufficiently hydrophilic to be excreted
efficiently by renal or biliary processes.

UDP-glucuronyltransferase (EC 2.4.1.17) is without doubt the most
important (and most widely studied) of these conjugating enzyme systems.
The specific substrate-donor molecule for glucuronidation is UDP-
glucuronate, and the glucuronate residue can be transferred to a wide
range of compounds, particularly those containing aryl -OH, -NH, or
-SH groups (ref. 1). The diversity of glucuronidation reactions has

required the development of a corresponding diversity of assay methods, according to the type of acceptor molecule employed (refs. 1 and 2). Generally, aliquots of the reaction mixture are sampled at intervals, and the residual acceptor concentration, or the product concentration, is determined following a solvent extraction procedure. A specific property of a component of the reaction (e.g. fluorescence or UV absorbance) can be exploited for the assay. Alternatively, radioactively labelled substances must be employed. In rare cases, an assay can be carried out with continuous monitoring, where photometric properties of acceptor and product differ, as for example with a chromophoric compound such as p-nitrophenol. The most widely examined class of acceptors are the simple phenols, and a general assay based on a colorimetric reaction with unconjugated acceptor has recently been reported (ref. 3).

The substrate for glucuronidation, UDP-glucuronate is not only metabolised by the conjugating enzyme systems in the microsomal fraction of mammalian liver. There is considerable hydrolytic activity towards this nucleotide, which is found in all but the purest preparations of UDP-glucuronyltransferase. Detergent-solubilisation procedures on the microsomal fraction result in both glucuronyltransferase and pyrophosphatase activities in the 100,000 g supernatant, and a considerable amount of further purification is necessary to separate the two activities. These two enzyme systems appear to contribute to the so-called glucuronic acid pathway, which is shown schematically in Fig. 1. Thus, UDPGA is employed in the biosynthesis of glucuronide conjugates via UDP-glucuronyltransferase (EC 2.4.1.17), which in turn can be hydrolysed by β-glucuronidase (EC 3.2.1.31), yielding free glucuronate (glucuronic acid). Alternatively, pyrophosphate cleavage of the UDPGA through a non-specific nucleotide pyrophosphatase (EC 3.6.1.9) gives glucuronate 1-phosphate, which through hydrolysis by a general alkaline phosphatase (EC 3.1.3.1) also results in the production of free glucuronic acid. The unspecific pyrophosphatase also catalyses the hydrolysis of UDPG, NAD^+, NADH, $NADP^+$, and NADPH, so that this pathway is probably also competitive with respect to other drug-metabolising systems, such as the monooxygenases which require nicotinamide dinucleotides.

To date, two basic methods have been reported for the assay of UDP-glucuronate hydrolysis. A complete analysis is only possible using

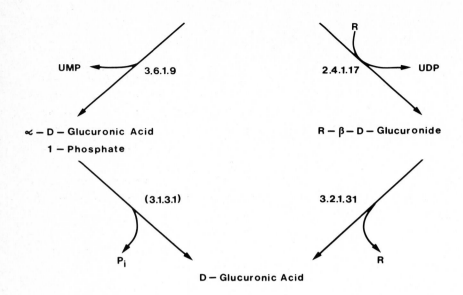

UDP — ∝ — D — Glucuronic Acid

UMP   3.6.1.9

R

2.4.1.17   UDP

∝ — D — Glucuronic Acid
1 — Phosphate

R — β — D — Glucuronide

(3.1.3.1)

3.2.1.31

P$_i$

R

D — Glucuronic Acid

Fig. 1. Schematic representation of the glucuronic acid pathways from UDP-glucuronate. The drug-metabolising route is via glucuronide synthesis, and cleavage by β-glucuronidase. The hydrolytic route is via glucuronate 1-phosphate, and further hydrolysis, probably by a non-specific alkaline phosphatase.

labelled UDPGA in combination with a paper- or thin-layer chromatographic separation (ref. 4). The metabolism of UDPGA via both pathways can be analysed using this method, but this advantage is somewhat reduced by the laborious nature of the analyses. A more rapid assay of the pyrophosphatase activity is possible in a coupled enzyme test using alkaline phosphatase, whereby the primary products UMP and GA-1-P are further hydrolysed to inorganic phosphate. The total phosphate liberated is estimated by the method of Fiske and SubbaRow, and the amount of UDPGA cleaved is directly calculated as half the total phosphate liberated (ref. 5). This method is unfortunately only applicable to the pyrophosphatase activity. Any inherent non-specific phosphatase activity in enzyme preparations is not detected.

The present contribution offers an alternative quantitative and simultaneous assay of nucleotide pyrophosphatase and glucuronyltransferase activities. The influence of further phosphatase activity is also detectable using this analysis. In capillary isotachophoretic separation, most of the components of the glucuronic acid pathway can be determined in one analysis under suitable conditions. Thus, at pH 3.89, the charged species UDPGA, UMP, GA-1-P, P_i, GA, UDP, glucuronides, can be separated and quantified. The only components of the total scheme which are not found in the tachopherogram are uridine, from complete hydrolysis of UMP, and the glucuronate acceptor, which can be analysed by alternative methods.(ref. 3).

MATERIALS AND METHODS

Instrumentation and Analysis

The isotachophoretic analyses were carried out on an LKB 2127 Tachophor (LKB-Produkter AB, Bromma 1, Sweden), fitted with a Teflon capillary of internal diameter 0.5 mm and length 61 cm. The capillary block was maintained at a temperature of 10°C during the measurements and the zones were detected photometrically at 254 nm.

The leading electrolyte system was 5 mmole/l HCl containing 0.25 % (w/v) HPMC (hydroxypropylmethylcellulose, Dow Chemical Co. Midland, Mich. U.S.A.) corrected to pH 3.89 by the addition of β-alanine. The terminating electrolyte was 5 mmole/l caproic acid.

Isotachophoretic separations were carried out under conditions of constant current - 100 µA for the first 20 min, and then at 50 µA. The total analysis time was of the order of 30 min, and the samples passed the detector at potentials in the range 17 to 20 kV.

Calibration with Standard Solutions

The reference substances for the analyses (UTP, UDP, UMP, UDPG, UDPGA, GA-1-P, and GA) were purchased from Boehringer (Mannheim, G.F.R.) Sigma (St. Louis,Mo., U.S.A.), and were of the highest available purity. Potassium dihydrogen phosphate and paracetamol (p-acetamidophenol) were obtained from Merck (Darmstadt, G.F.R.), and were of analytical quality. TRIS.HCl buffers for enzyme assays were made using Trizma Base (Sigma).

Solutions of the reference substances were prapared to a concentra-tion of 1 mmole/l in 100 mmole/l TRIS.HCl buffer (pH 8) as standards for the calibration of zone lengths. Between 1 and 10 nmole of each of these substances were injected into the instrument in a volume no greater than 10 µl, and calibration plots were derived from the zone lengths in this range. A number of small UV-absorbing and non-absorbing zones were detected from impurities in the electrolyte systems; these were differentiated from impurities in the reference compounds from a separation in the absence of added sample.

The order of mobilities of the reference samples was deduced both from the relative heights of the thermal signals of the individual samples, and from the order of separation of successive mixtures of the samples.

Enzyme Assays

The microsomal fraction of homogenates of rat, rabbit and pig liver were prepared by differential centrifugation. Protein determinations were carried out by the method of Lowry et al. All enzyme assays were carried out at 25°C in 100 mmole/l TRIS.HCl buffer pH 8.

Nucleotide pyrophosphatase was assayed with an initial UDPGA concentration of 0.5 mmole/l, and incubations were carried out in the absence and presence of 1 mmole $MgCl_2$/l and 2 mmole EDTA/l to establish the divalent cation dependence of the enzyme. The protein concentration in the assay mixture was adjusted to about 1 mg/ml. Immediately after initiation of the reaction by addition of enzyme to the incubation mixture, a 10 µl aliquot was removed, and injected into the Tachophor. The separation was carried out under the conditions described above. Further 10 µl portions of the mixture were analysed at intervals of 30 to 40 min until no more UDPGA was detectable on the tachopherogram. To ensure that the reaction was quenched on injection into the instrument, a parallel incubation was carried out, and aliquots were removed at similar time intervals. In this case, reaction was stopped by addition of a tenth of the sample volume of 70% perchloric acid to precipitate the protein. At the completion of reaction, the samples were analysed on the Tachophor. Both methods gave similar results, suggesting that direct injection into the Tachophor provides satisfactory quenching of the reaction. The acid precipitation method, however, is more suitable for reactions in which aliquots must be analysed at time intervals of less than the separation time in the Tachophor (about 30-40 min).

UDP-glucuronlytransferase was assayed with initial UDPGA and $MgCl_2$ concentrations of 2.5 and 1 mmole/l, respectively, and with a total protein content of approximately 1 mg/ml. The p-acetamidophenol concentration was varied up to a maximum concentration of 2 mmole/l. Aliquots of the reaction mixture were removed and analysed on the Tachophor as for the nucleotide pyrophosphatase. Under the conditions employed, the glucuronidation of p-acetamidophenol is quite slow, so that a time interval of 30 to 40 min (the time required for the development of a tachopherogram) was found to be suitable for sampling the mixture. More rapid reactions could be analysed by shorter sampling intervals, and quenching of the process by precipitation with 70 % perchloric acid as described above.

In order to provide a reference method for the isotachophoretic assay of glucuronidation, the colorimetric method using Folin's reagent as described recently by our group (ref. 3) was employed. Aliquots (0.2 ml) of the reaction mixture were sampled at regular time intervals. The reaction was quenched by the addition of 0.02 ml of 70 % perchloric acid. After centrifugation, 0.1 ml of the clear supernatant was removed, and diluted to a total volume of 1 ml with distilled water. A 0.1 ml volume of Folin-Ciocalteu reagent (Merck), and 1 ml of 10 % sodium carbonate solution were then added, and the mixture was left to stand at room temperature for 1 hour. The absorbance of the solution was measured at 750 nm against a sample prepared in a similar way from 0.1 ml of 100 mmole/l TRIS.HCl (pH 8) in place of the reaction sample. The unreacted p-acetamidophenol was calculated from a calibration curve drawn from the absorbances of standard solutions.

RESULTS

The order of migration of the uridine nucleotides under the analytical conditions described in Materials and methods is shown in the tachopherogram in Fig. 2. UDPGA has the highest ionic mobility, followed by UDP, UDPG and UMP, as the five UV-absorbing zones, a, b, c, e, and g. The non-absorbing zones d and f represent inorganic phosphate and glucuronate ions, respectively. Each of the ionic species was present to the extent of 1 to 2 nmole. Two small peaks are seen, and are due to impurities in the buffer systems.

Fig. 3 depicts the principles of analysis of zones of a tachopherogram with a UV-absorbing and non-absorbing substance. A prerequisite for the accurate determination of these zones is symmetry of the peak.

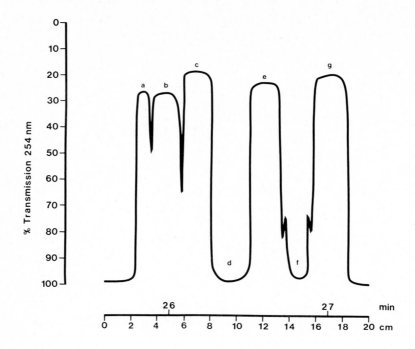

Fig. 2. Tachopherogram of a mixture of uridine nucleotides at pH 3.89.
UV-absorbing zones a, b, c, e, and g are due to UDPGA, UTP,
UDP, UDPG, and UMP, respectively. The non-absorbing zones d
and f are due to inorganic phosphate and glucuronate, respectively. Each ionic species was present to the extent of 1 to
2 nmole. Glucuronate 1-phosphate was not present in this mixture, but migrates under these conditions between the UDPGA
and UDP zones b and c. Two small UV-absorbing peaks are seen,
and are due to impurities in the buffer systems.

This method was employed both for the calibration experiments, and for
the enzymatic reactions. The width of the UV-absorbing zone, D_x, was
measured at half its height, H_x, and was converted into units of time
according to the set chart speed. The width of a non-absorbing zone,
D_y, was estimated at half the height, H_y, of the lower neighbouring
absorbing peak, which can either be another sample zone, or a peak
due to an impurity. Calibration plots for the species involved in the
enzymatic processes described here were constructed as zone widths in
seconds with respect to the absolute amount of sample injected in

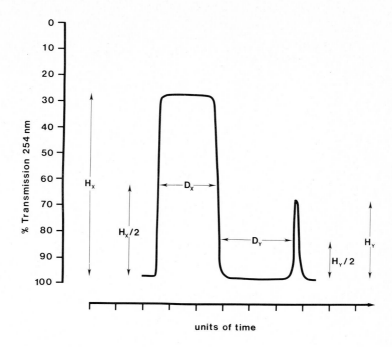

Fig. 3. Method employed for evaluating the zones of a tachopherogram. A UV-absorbing zone is taken as the width of the zone D_x at half the peak height H_x. A non-absorbing zone is taken as the width D_y at half the height of the lower neighbouring UV-absorbing peak H_y.

nmoles. As shown in Fig. 4 and 5, these plots were all linear within the range studied.

The tachopherogram shown in Fig 6 are representative of the results of incubations of UDPGA with microsomes derived from rabbit liver homogenate. The concentrations of reaction components for the study of UDPGA hydrolysis were as defined in materials and methods. The upper tachopherogram shows the composition of the incubation medium at an early stage of the reaction. The major component is still the substrate UDPGA (zone a), with a certain amount of UMP produced (zone i). Apart from the non-UV-absorbing phosphate zone (zone d), the other products are only present at minimal concentrations. At a later stage of the reaction, as shown in the lower tachopherogram, the amount of UDPGA is considerably reduced. However, the UMP zone (zone i) has not

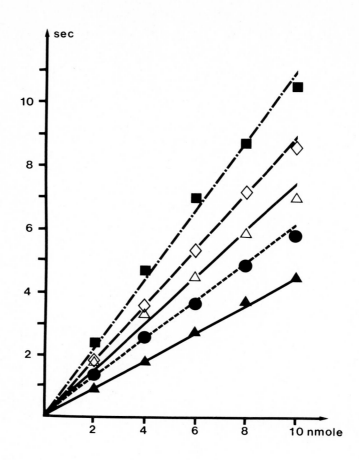

Fig. 4. Calibration plots for UDPGA (■), UDP (◇), UMP (△), GA (●), and GA-1-P (▲), expressed in terms of the absolute amount of reference sample applied to the Tachophor in nmoles. The width of each zone was calculated as described in the text, and converted to units of time according to the set chart speed.

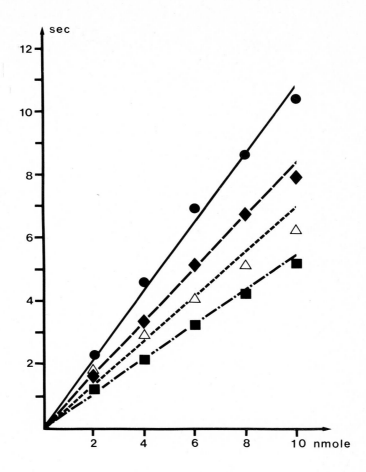

Fig. 5. Calibration plots for UDPG (●), UTP (◆), G-1-P (△), and P$_i$ (■), expressed in terms of the absolute amount of reference sample applied to the Tachophor in nmoles. The width of each zone was calculated as described in the text, and converted to units of time according to the set chart speed.

164

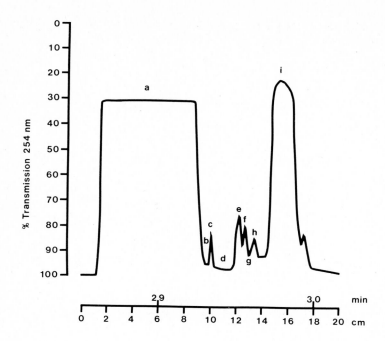

Fig. 6. Tachopherograms showing stages of hydrolysis of UDPGA by nucleotide pyrophosphatase. The upper diagram is at an early stage of the reaction, and the lower diagram is at a considerably later stage. Peaks a and i are due to UDPGA and UMP, respectively. The non-absorbing zones b, d, and g are due to GA-1-P, inorganic phosphate, and GA, respectively. Impurities in the system are shown by the small peaks, c, e, f, and h. The chart speed employed was 6 cm/min.

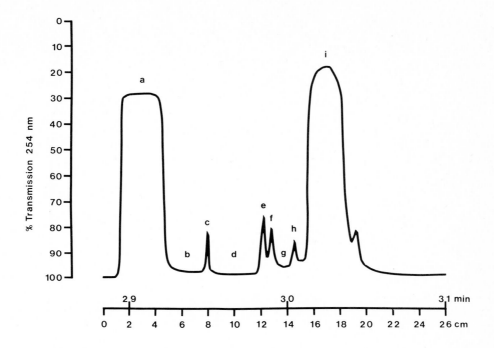

increased proportionately. Thus, the total uridine nucleotide content of the incubation has decreased. This must mean that UMP is itself further hydrolysed to the nucleoside, which is not detected on the tachopherogram under these conditions. The validity of this assumption is verified by a study of the fate of the carbohydrate residue of UDPGA. A zone of increasing width due to glucuronate 1-phosphate is observed (zone b), together with an increasing zone due to glucuronate itself,(zone g), demonstrating that a further hydrolytic activity, namely a phosphatase, is present in the microsomal preparation. Furthermore, the phosphate (zone d) increases at a rate which cannot be explained solely by the final hydrolysis of the GA-1-P. The other source of phosphate must be via hydrolysis of UMP to the nucleoside. The kinetics of this system are shown diagrammatically in Fig. 7. Initially, the UDPGA concentration decreases linearly with time (i.e. over the first 100 min of reaction). During this period, the UMP concentration is more or less constant, showing a steady-state situation in which the production of UMP by pyrophosphatase cleavage of

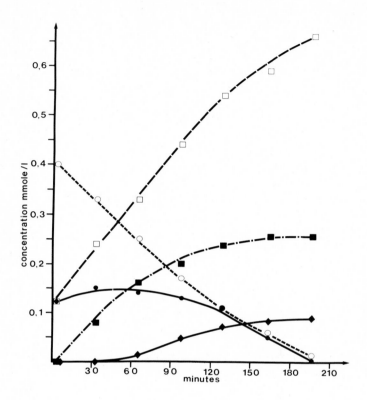

Fig. 7. Kinetic profile of the hydrolysis of UDPGA obtained from
isotachophoretic data. The initial substrate, UDPGA (○),
is hydrolysed within about 200 min. UMP concentrations (●)
are more or less constant during the first 40% of reaction
time. The other primary product of UDPGA hydrolysis, GA-1-P
(■), displays an increasing concentration for the first
2 h, and levels off when the further hydrolysis to GA (◆)
becomes appreciable. The other final hydrolysis product,
inorganic phosphate (□), increases linearly during the
first 2 h of reaction.

UDPGA approximatively balances the further phosphatase cleavage of
UMP to uridine and inorganic phosphate. Glucuronate 1-phosphate increases with decreasing UDPGA concentration, again quite linearly

until the UDPGA is considerably depleted. The product of GA-1-P
hydrolysis, glucuronate, only begins to be produced to a significant
extent when the GA-1-P concentration has reached a reasonable level.
The total inorganic phosphate concentration rise is at a high rate,
corresponding to its production from two sources, namely from the
phosphatase cleavage of both GA-1-P and UMP. It is possible to resolve
the phosphate components, denoted by P(g) and P(u) (derived from
GA-1-P and UMP, respectively), from the total phosphate P(t) determined from the isotachophoretic zone. A further quantity P(o) must
be defined to take account of the trace of inorganic phosphate initially present as impurity in the incubation system.

Thus $\qquad$ P(t) = P(g) + P(u) + P(o)

The component P(g) must be identical with the concentration of GA,
the other product of GA-1-P hydrolysis, so that the expression can
be rearranged as

$\qquad$ P(u) = P(t) - P(o) - GA

By a similar argument, P(u) is equal to the uridine concentration,
being the two equiproportional products of UMP hydrolysis.

Thus, the UDPGA concentration at a particular stage of the reaction
can be estimated in three ways: (a) directly from the width of the
UDPGA zone, (b) from the sum of the GA-1-P and GA concentrations
deducted from the initial UDPGA concentration, and (c) from the sum
of UMP and P(u) concentrations deducted from the initial UDPGA concentrations. Of course methods b and c are both dependent on the
concentration of GA, method b directly, and method c indirectly.

Table I summarises the isotachophoretic analysis of the reaction
shown in Fig. 7. The concentrations are derived from the zone widths,
quantified from the calibration curves. The quantities P(g) anf P(u)
are calculated indirectly as described above. The three calculations
of the UDPGA concentrations during the course of the reaction are
summarised in Table II. The values derived from the GA-1-P and GA
concentrations are in good agreement with the direct measurements
of UDPGA at the later stages of hydrolysis. During the early stages
of the reaction, the amounts of the products generated are probably

TABLE I

Concentrations of UDPGA, GA-1-P, GA, UMP, and total phosphate, determined directly from the widths of the zones at the given time intervals.
The phosphate component P(g) is taken to be equal to the GA concentration, and P(u) is derived indirectly as described in the text.
The initial UDPGA concentration was 0.5 mmole/l in this experiment, and the initial phosphate concentration P(o) was 0.12 mmole/l.

Time (min)	Concentration (mmole/l)					
	UDPGA	GA-1-P	GA = P(g)	UMP	P(t)	P(u)
33	0.33	0.08	0.00	0.15	0.24	0.12
65	0.25	0.16	0.03	0.14	0.33	0.18
97	0.17	0.20	0.10	0.13	0.44	0.22
129	0.11	0.24	0.15	0.12	0.54	0.27
163	0.06	0.26	0.17	0.05	0.59	0.30
196	0.03	0.26	0.18	0.00	0.66	0.36

too small to be deduced accurately. Thus, the quantity deducted from the initial UDPGA concentration (GA-1-P + GA) is underestimated, and the resulting value for the UDPGA concentration is too high. On the other hand, P(u) values are also dependent on the deduction of the GA concentrations. The P(u) values are thus overestimated at early stages of the reactions, and the resulting UDPGA concentrations derived are too high. This effect can be cancelled out by calculating a mean value for UDPGA from the two indirect estimations. These values were found to be in very good agreement with the direct measurements of UDPGA concentrations.

A lineweaver-Burk plot for rabbit liver microsomal UDPGA-pyrophosphatase is shown for varying initial UDPGA concentrations in Fig. 8. The Michaelis constant from the plot was 0.36 mmole/l, which is of the same order of magnitude as that found previously (ref. 5) with rat liver microsomal enzyme.

Liver microsomes from all three mammalian sources, rabbit, rat, and pig, were found to contain high activities both of pyrophosphatase and phosphatase. These reactions, with UDPGA, UDP, UMP, or GA-1-P, were all found to be higher in the presence of magnesium ions, which is in agreement with previous reports in the literature.

TABLE II

The concentrations of UDPGA during the course of the hydrolysis,
estimated according to three methods from the data of Table I.
Method A is from the direct analysis of the UDPGA zone. Method B is
derived from the sum of the GA-1-P and GA concentrations deducted
from the initial UDPGA ocncentration, and method C from the sum of
the UMP and P(u) concentrations deducted from the initial UDPGA
concentration as explained in the text. The values in column D are
the mean values from Methods B and C.

Time (min)	UDPGA Method A (mmol/l)	UDPGA Method B (mmol/l)	UDPGA Method C (mmol/l)	UDPGA D (mean of B and C) (mmol/l)
33	0.33	0.42	0.23	0.33
65	0.25	0.31	0.18	0.25
97	0.19	0.20	0.15	0.17
129	0.11	0.11	0.11	0.11
163	0.06	0.07	0.15	0.11
196	0.03	0.06	0.14	0.10

Solubilisation of the microsomal fraction by the action of detergents or lipase resulted in the recovery of the majority of these
activities in the 100,000 g supernatant. Thus, effects due to pyrophosphate cleavage of UDPGA would be expected to disturb measurements
of UDP-glucuronyltransferase, which also requires UDPGA as substrate.
This was indeed found to be the case, such that glucuronidation could
only be studied satisfactorily in enzyme preparations which had been
subjected to further purification procedures. We therefore partially
purified a UDP-glucuronyltransferase according to methods described
previously (ref. 6) for the purposes of this study.

The acceptor for glucuronidation chosen was p-acetamidophenol,
which as paracetamol is an analgesic drug of considerable current
interest. There were two main reasons for choosing this substrate
for isotachophoretic studies. Firstly, paracetamol is glucuronidated
at a rate considerably less than other simple phenols. Thus the effects
due to donor molecule hydrolysis become of great importance. Furthermore, paracetamol glucuronide has a much lower mobility under our
analytical conditions than other components of the reaction mixture.

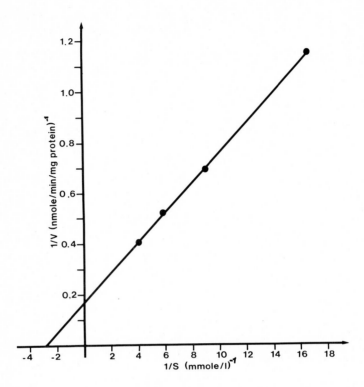

Fig. 8. A Lineweaver-Burk plot showing reciprocal initial rate activities of UDPGA pyrophosphatase activity with respect to
reciprocal initial UDPGA concentrations. The Michaelis constant
from the plot was found to be 0.36 mmole/1 from the intercept
on the x-axis.

Fig. 9 shows a tachopherogram taken during the course of glucuronidation of paracetamol. The UV-absorbing zones a and f are due to
UDPGA and UMP, respectively. The new peak g is due to paracetamol
glucuronide. The non-absorbing zones c and e are due to inorganic
phosphatae and GA, respectively. It is interesting to note that
virtually no UDP was detected on the tachopherogram, although this

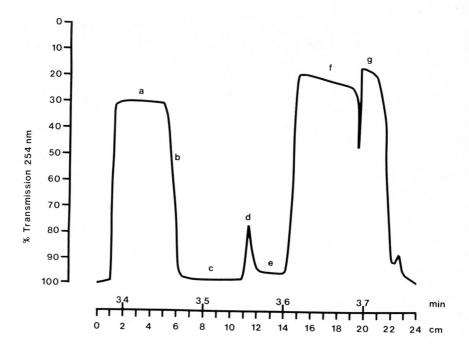

Fig. 9. Tachopherogram from incubation of p-acetamidophenol (2 mmole/l)
with UDPGA (2.5 mmole/l) in the presence of magnesium ions
(1 mmole/l), with a partially purified cholate-solubilised preparation of UDP-glucuronyltransferase in 100 mmole/l TRIS.HCl
buffer (pH 8) at 25°C. The zones detected were: UDPGA (a),
UDP as a shoulder (b), inorganic phosphate (c), an impurity
in the system (d), GA (e), UMP (f), and paracetamol glucuronide
(g). This represents the stage of reaction after about 300
minutes of incubation.

should be a primary product of glucuronidation. In fact, the small
shoulder b on the UDPGA zone is due to this nucleotide. Thus, there
is considerable general phosphatase activity even in this partially
purified enzyme, which immediately cleaves any UDP formed. This was
confirmed by incubating UDP with the enzyme; all the UDP was hydrolysed
within a few minutes. The activity towards UMP was considerably lower.
A certain amount of GA was detected on the tachopherogram, although
GA-1-P was not found at any stage of the reaction. Thus, it would
seem that our purification procedures result in the elimination of
most of the pyrophosphatase activity, so that this reaction becomes
rate-limiting, but the general phosphatase activity remains high.

The unpurified enzyme from microsomes would have hydrolysed all the UDPGA in the incubation mixture within the 300 min of reaction time, as for the tachopherogram shown in Fig. 9.

Fig. 10 represents the kinetic profile of the reaction of which Fig. 9 was a typical tachopherogram. In this case, there are two possible routes of reaction for UDPGA. Firstly, glucuronidation results in the production of paracetamol glucuronide. This pathway can be analysed directly from the increasing widths of the glucuronide zone. The other product of this reaction is UDP, which is quantitatively hydrolysed to UMP in our system. The alternative route is via UDPGA cleavage by pyrophosphatase, as described earlier. Since we found little or no GA-1-P on the tachopherogram, we must assume that this activity can be quantified by the conversion to UMP and GA, the latter as a final product. We can not, unfortunately, assess the further hydrolysis of UMP to U and inorganic phosphate in this case, since uridine is not detectable on the tachopherogram under these conditions. However, since pyrophosphatase activity is rate-limiting for this second route of UDPGA usage, we can deduce the amount of UDPGA metabolised in the hydrolytic pathway directly from the amount of GA generated. Thus, deduction of the GA concentration from the initial UDPGA concentration yields the quantity of UDPGA utilised in hydrolysis. The difference between this value and the experimentally measured amount of UDPGA remaining in the mixture is a direct reflection of the amount of UDPGA utilised in glucuronidation. This quantity should be identical with the amount of paracetamol glucuronide formed.

In these experiments, we required a reference method for the glucuronidation reaction. We therefore employed the colorimetric method described previously by our group (ref. 3). Here, the unreacted paracetamol is assessed by an analysis with Folin-Ciocalteu reagent. Fig 11 shows the initial rate enzyme kinetics for paracetamol glucuronidation determined from the three methods described above. Thus, the increasing paracetamol glucuronide concentration was determined directly from the relevant zone on the tachopherogram, from the reference method with Folin-Ciocalteu reagent, and from a calculation of the amount of UDPGA employed in glucuronidation by correction for the hydrolytic pathway. As seen in Fig. 11, these three methods yield a more or less identical activity for the UDP-glucuronyltransferase.

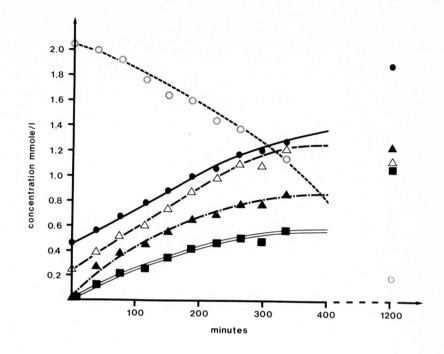

Fig. 10. Kinetic profile for the glucuronidation of p-acetamidophenol, parallel to UDPGA hydrolysis. Quantitative estimation of enzyme activities can be obtained from the rates of increase in product concentrations, inorganic phosphate (●), UMP (△), GA (▲), and paracetamol glucuronide (■), or from the decrease in UDPGA (◎), as described in detail in the text.

DISCUSSION AND CONCLUDING REMARKS

It has been shown in this contribution that analytical isotacho-phoresis can be applied to the study of enzyme systems employing the nucleotide UDP-glucuronic acid as substrate. The method can equally well be applied to reactions with other nucleotides, provided that analytical conditions can be selected to give adequate separation of the reactants and products of the process.

UDPGA is the substrate for two processes in liver microsomes, firstly as substrate donor molecule for glucuronidation reactions, but also for the hydrolytic nucleotide pyrophosphatase. At least _in vitro_ these enzymes compete for UDPGA as a substrate, but in unpurified

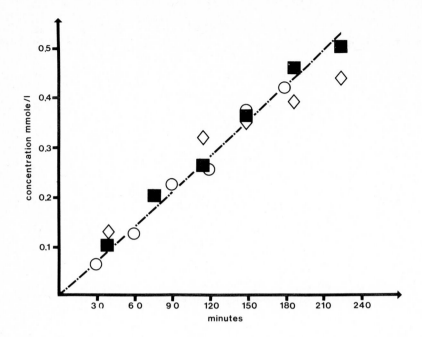

Fig. 11. The increase in concentration of p-acetamidophenylglucuronide
during the course of 240 min of reaction monitoring. The
values were obtained in three ways as described in detail in
the text: (1) The free phenolic concentration was determined
by a colorimetric method, and the result was deducted from
the initial concentration (shown by◯). (2) The sum of the
remaining UDPGA concentration and GA concentration was deducted from the initial UDPGA concentration to deduce the amount
of UDPGA employed in glucuronidation (shown by◇). (3) The
glucuronide zone was determindd directly from the tachopherograms (shown by■).

preparations the activity of the hydrolic pathway exceeds that of
glucuronidation by a factor of at least 10. This has serious conse-
quences for workers interested in accurate assessments of UDPGA effects
in the glucuronidation system. Even in partially purified enzyme
preparations from solubilised microsomal protein, hydrolytic activities
are high enough to lead to a misinterpretation of glucuronidation
results. Isotachophoresis provides an excellent method for the complete
analysis of such systems. The sensitivity of the method is more than
adequate for the use of concentrations normally employed in biochemical

studies. Further advantages of the method are that no radioactively labelled substances are required as for thin-layer chromatographic techniques, the analysis is rapid, and requires very little sample.

It is reasonable to predict that analytical capillary isotachophoresis will find considerable applications to a wide range of enzymological studies in the future, particularly where nucleotide substrates are concerned. However, this process will be hindered by inborn prejudices, conservatism towards more classical, but not necessarily better, analytical methods. Moreover, the efforts of those interested in the use of such revolutionary methods will be hampered by general ignorance in the field, and by the stubbornness of incompetent referees.

ACKNOWLEDGMENTS

The author has received vital help from many sources, for which he is most deeply grateful. First and foremost, thanks are due to Mrs. Sabine Husmann-Holloway for doing all the work described in this paper. A special acknowledgment is also due to Dr. Gorig Brunner, in whose laboratory the experiments were carried out, support by grants from the Deutsche Forschungsgemeinschaft. Without the interest shown, and help afforded by Mr. Astor Baldesten and his collegues at L.K.B. this research would not have been possible. The general advice and encouragement of Professors I. Trautchold, E. Schmidt, and F.W. Schmidt is also gratefully acknowledged.

REFERENCES

1 G.J. Dutton in G.J. Dutton (Editor) Glucuronic Acid, Free and combined, Academic Press, New York, (1966), PP 280-286.
2 G.J. Dutton and B. Burchell, in J.W. Bridges and L.F. Chasseaud (Editors) Progress in Drug Metabolism, Wiley-Interscience, London, 1975, PP 1-70.
3 C.J. Holloway, S. Husmann-Holloway and G. Brunner, Z. Naturforsch. (1979) in press.
4 E. Puhakainen and O. Hänninen, Eur. J. Biochem. 61(1976)165-169.
5 E. Puhakainen, A. Saarinen and O. Hänninen, Acta Chem. Scand. B 31(1977)125-129.
6 G. Brunner, C.J. Holloway and H. Lösgen, Int. J. Artif. Organs, (1979) in press.

A. Adam and C. Schots (Editors), *Biochemical and Biological Applications of Isotachophoresis*   177
© 1980 Elsevier Scientific Publishing Company, Amsterdam — Printed in The Netherlands

THE ISOTACHOPHORETIC ANALYSIS OF β-GLUCURONIDES: A NEW APPLICATION
IN THE ENZYMATIC ASSAY OF UDP-GLUCURONYLTRANSFERASE

C.J. Holloway, S. Husmann-Holloway and G. Brunner
Institut für Klinische Biochemie, Abteilung für Gastroenterologie,
Medizinische Hochschule Hannover, Karl-Wiechert-Allee 9,
D3000-Hannover 61 (G.F.R.)

UDP-glucuronyltransferase (EC 2.4.1.17) is an important enzyme for
hepatic detoxification functions. The donor molecule (substrate) is
UDP-glucuronic acid, and the mechanism of detoxification is such that
the hydrophilic glucuronate moiety can be transferred to a wide range
of hydrophobic toxins containing a suitable reactive group such as
-OH, -NH$_2$, -COOH, etc. The product, a so-called β-glucuronide, is thus
far less hydrophobic than the original toxin, the glucuronate acceptor,
and can accordingly be excreted quite efficiently by renal or biliary
processes. The competition between two microsomal enzyme systems for
UDP-glucuronate as substrate has recently aroused some interest, since
the regulation of detoxification with respect to other essential
metabolic pathways must be involved. Such studies, have from the start,
been faced with considerable analytical difficulties. Reports from
our research group have shown that analytical capillary isotachophoresis
can be applied to a simultaneous study of UDP-glucuronate pyrophosphatase and UDP-glucuronate glucuronyltransferase activities. The main
advantage of isotachophoretic analysis, however, lies in the assay
of glucuronyltransferase, due to the simplicity of analysing the
products of glucuronidation, the β-glucuronides. Since many different
classes of acceptor can be glucuronidated, a diversity of analytical
techniques has been developed for the study of a particular acceptor
of interest. Analytical isotachophoresis is not as limited in its
application as, for example, photometric methods. For the assay of
glucuronyltransferase, it is as versatile as chromatographic techniques,
but is simpler to use, and is more accurate.

Another contribution in these proceedings describes the principles
of the assay of enzymatic processes involving UDP-glucuronate as

substrate using analytical capillary isotachophoresis. Our preliminary
work concerned the glucuronidation of paracetamol (p-acetamidophenol),
a drug of current importance, particularly in studies of liver failure.
The isotachopherogram in Fig. 1 shows that a complete analysis of
enzyme assays of the glucuronidation of this acceptor is possible, both
from the point of view of increasing product zones, and of decreasing
substrate zone. In fact, a great deal of boundary information can be
gained from the study of a single isotachopherogram. Let us study
Fig. 1 in some detail. The incubation mixture originally contained
UDP-glucuronate as donor molecule, and paracetamol as acceptor. The
former is seen by the UV-absorbing zone "a". The acceptor is not a
component of the isotachopherogram under these conditions. In principle,
the course of glucuronidation should result in a decrease in the width
of zone "a", as is indeed observed, but also in two other zones of
increasing width, namely for the products UDP and paracetamol glucuronide. UDP should appear at the position marked "b", but at the most
a small shoulder is seen. The glucuronide is seen as a UV-absorbing
zone "g". The fact that a further zone "f" is found to be UMP might
lead to the rapid assumption that the UDP has been hydrolysed by the
rather impure enzyme preparation. This proves to be true, but does
not explain why more UMP is found than paracetamol glucuronide when,
in principle, a ratio of 1:1 is expected from the glucuronidation
reaction. The solution is found by consideration of the non-UV-absorbing zones "c" and "e", attributable to inorganic phosphate, and
glucuronate, respectively. The phosphate arises, at least in part,
from the hydrolysis of UDP. The glucuronate, however, can only have
originated from the substrate UDP-glucuronate itself, and we can
immediately conclude that our enzyme preparation contains an activity
which is hydrolytic towards UDP-glucuronate. It is known that a pyrophosphatase activity is present in the same organelles as UDP-glucuronyltransferase. The initial products of this enzymatic hydrolysis are
glucuronate 1-phosphate and UMP. A non-specific phosphatase activity
would further cleave these compounds to glucuronate, phosphate, and
uridine. The "oversized" UMP zone is thus explained. This illustration
demonstrates the potential of analytical isotachophoresis when applied
to such problems as are discussed here.

We can now extend this work to cover a range of acceptors other
than paracetamol, and even to the study of the simultaneous glucuronidation of mixtures of acceptors. In this study, we have primarily

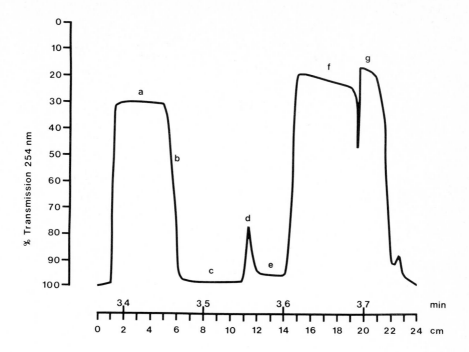

Fig. 1. Isotachopherogram from an incubation of paracetamol (p-acetamidophenol), 2 mmol/l, with UDP-glucuronate, 2.5 mmole/l, and
magnesium ions, 1 mmole/l, with microsomal UDP-glucuronyltransferase from rabbit liver. The reaction was carried out in 100
mmol/l Tris/HCl buffer (pH 8) at 25°C. The zones detected were:
UDP-glucuronate "a"; UDP, shoulder "b"; phosphate, "c"; an
impurity from the electrolyte systems, "d"; glucuronate, "e";
UMP, "f"; and the β-glucuronide, "g".

concerned ourselves with acceptors containing a phenolic hydroxyl
group.

The β-glucuronides were synthesised enzymatically from an incubation
mixture containing the respective acceptor (phenol, naphthol, p-nitrophenol , etc.) at a concentration of 2 mmol/l, and the donor molecule,
UDP-glucuronate, at a concentration of 5 mmol/l . Optimal activity was
achieved by the presence of magnesium ions at a concentration of 2
mmol/l. The enzyme preparation was the microsomal fraction of rabbit
liver homogenate, prepared by differential centrifugation. The course
of the glucuronidation reaction was monitored by the decrease in free
acceptor concentration, employing a photometric test with Folin-

Ciocalteu reagent which gives a colour reaction only with the unconjugated phenol. Among the substrates used in these studies were the
1- and 2-isomers of naphthol. The enzymatic reaction for these two
acceptors is shown in Fig. 2. Under the identical conditions employed,
it is seen that 1-naphthol is glucuronidated considerably more rapidly
than 2-naphthol. Generally, the reaction was found to be complete
within 6 h, with a microsomal protein concentration of 1 mg/ml in the
incubation. The mixture was left to stand for a further 4 h, which
resulted in hydrolysis of the residual UDP-glucuronate through pyrophosphatase activity in the microsomal protein. Upon completion of
the incubation, the mixtures were centrifuged at 100,000 x g to remove
the enzyme. The supernatant thus contained the phenolic β-glucuronide
at a concentration of 2 mmol/l. This solution was employed for the
calibration experiments in the isotachophoretic equipment.

The analytical conditions for the isotachophoretic measurements
were as follows: The LKB 2127 Tachophor was used with a PTFE capillary
of internal diameter 0.5 mm and length 61 cm. The system was thermostatted at 10°C, and the zones were detected photometrically at 254nm.

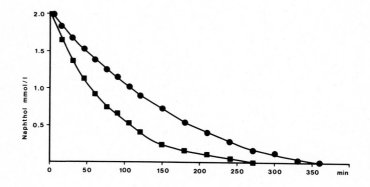

Fig. 2. Kinetics of glucuronidation of 1-naphthol (squares) and
        2-naphthol (circles) monitored as the decrease in free phenolic
        concentration with Folin-Ciocalteu reagent. Under identical
        conditions, the 1-naphthol is glucuronidated more rapidly than
        the 2-isomer.

The leading electrolyte system was 10 mmol/l HCl corrected to the
desired pH by the addition of β-alanine. This system also contained
HPMC at a concentration of 0.25 %. The terminating system was hexanoic
acid, 10 mmol/l. An initial current of 100 μA was applied, and was
reduced to 50 μA shortly before completion of the separation. The
total time for analysis was of the order of 30 to 45 min, and the
samples passed the detector at potentials of 17 to 20 kV.

β-Glucuronides were synthesised from p-nitrophenol, phenol, 1-
naphthol, 2-naphthol, phenolphthalein and paracetamol. The optimal pH
for the leading electrolyte system from the point of view of separation
of the nucleotides involved in glucuronidation has been found to be
around 3.9, but under such conditions the glucuronides tend to form
zones which are partially mixed with the UMP zone. The exception is
paracetamol glucuronide, whose mobility appears to be sufficiently
lower than that of UMP to permit a separation at pH 3.9. For all the
other phenylglucuronides it was found that a consistently better
separation from UMP was effected with decreasing pH, the optimum being
around pH 3.3. Naturally, the separations in regions of higher mobility
(i.e. around the zones for UDP-glucuronate or phosphate) are less well-
defined. A reasonable compromise can be taken at pH 3.5, but specific
conditions will have to be selected for the particular combination of
glucuronidation reactions to be studied in future work. At pH 3?3,
for example, p-nitrophenylglucuronide and phenylglucuronide formed a
mixed zone, of higher mobility than the naphthylglucuronides. The
isomeric forms of the naphthylglucuronides, however, formed a mixed
zone, which in turn migrated in front of the phenolphtalein glucuronide zone. The paracetamol glucuronide zone was the last in order of
migration. Thus it is possible to separate some, but not all, of the
glucuronides under uniform conditions. There would, therefore, be no
problems associated with the monitoring by isotachophoresis of the
simultaneous glucuronidation of a mixture of, say, phenol + 1-naphthol
+ phenolphthalein + paracetamol. For each product a distinct zone
would develop.

A rather more complex problem is faced when a mixed zone is formed
as, for example, with the naphthylglucuronides. The solution of this
problem resulted in the development of a rather unique exploitation of
zone lengths, and heights. In Fig. 2 the course of glucuronidation
of the isomers of naphthol is given, as determined by the decrease in

free naphthol concentration from photometric analysis. The isotachophoretic analysis of the increase in product from the naphthylglucuronide zone width is given in Fig. 3. Both methods gave similar enzymatic activities. A fascinating phenomenon was found in the construction of calibration plots of zone lengths with respect to the injected amount of naphthylglucuronide. The slopes of the plots were identical for the 1- and 2-isomers, demonstrating an identical "concentrating effect" for the two glucuronides. This is shown in Fig. 4. This means that at least the <u>total</u> amount of naphthylglucuronide in the zone can evaluated from the width of the zone, irrespective of the proportions of the individual isomers in the mixture. From the width alone, therefore the composition of the mixture cannot be deduced. As shown in Fig. 5, however, the molar extinction coefficients of the naphthylglucuronide isomers differ considerably. 2-Naphthylglucuronide absorbs more strongly at 254 nm than 1-naphthylglucuronide, and mixtures of the two isomers absorb within the range of the pure-isomer zones. The actual absorbance of the zone was found to be linearly dependant on the proportion of each isomer in the zone. Fig. 6 is a diagrammatic summary of

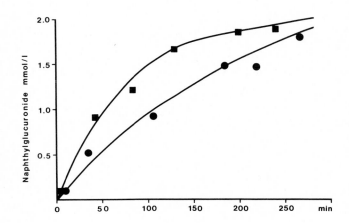

Fig. 3. The glucuronidation reactions depicted in Fig. 2, but monitored isotachophoretically by the increase in width of the glucuronide zones. The squares represent the values for 1-naphthylglucuronide, which is formed more rapidly than the 2-naphthylglucuronide shown by the circles. The pH of the leading electrolyte in the Tachophor was 3.3.

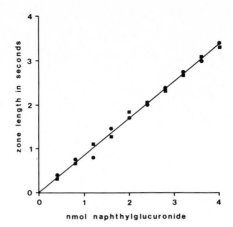

Fig. 4. Calibration curves from the isotachophoretic data for 1- and
2-naphthylglucuronides. The zone width, in seconds, is given
as a function of the amount of substance injected. The slope
of the plots is the same for 1-naphthylglucuronide (squares)
and 2-naphthylglucuronide (circles).

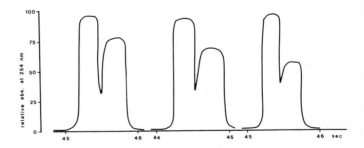

Fig. 5. Parts of isotachopherograms containing 1-naphthylglucuronide
(right-hand side) and 2-naphthylglucuronide (left-hand side),
and a mixture of the two isomers (middle). The more strongly
absorbing zone is UMP, which is constant in all cases. The 2-
2-isomer is seen to absorb more strongly at 254 nm than the
1-isomer.

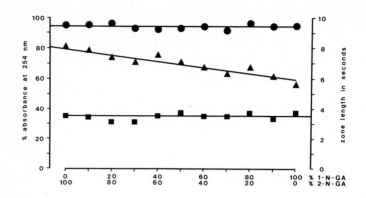

Fig. 6. Results of isotachophoretic analyses of 1-naphthylglucuronide
(1-N-GA) and 2-naphthylglucuronide (2-N-GA), and of mixtures
of the two isomers. The total amount injected was 5 nmole in
all cases, and the zone length is independent of the proportion
of each isomer in the mixture (squares). The height of the
UMP zome is constant (circles), demonstrating the stability
of the photometric detector. The height of the naphthylglucuronide zone is seen to depend linearly on the proportion of 1-
and 2-isomer in the mixture (triangles).

these observations. Five microlitres of a 2 mmol/l solution of naphthylglucuronide, with varying proportions of each isomer, were injected
into the Tachophor. The zone length was around 2.25 sec for all the
samples. In order to test the stability of the system with respect to
absorbance measurements, the height of the UMP zone was also determined
and found to be consistent at around 96 %. Pure 2-naphthylglucuronide
zones gave an absorbance at around 80 %, whereas the pure 1-naphthyl-
glucuronide zones absorbed less strongly, at around 60 %. A linear
decrease in absorbance at 254 nm was observed with decreasing proportion of 2-naphthylglucuronide in a mixed zone. Whereas the zone width
gives directly the total amount of naphthylglucuronide applied to the
instrument, the height of the zone gives the proportion of each isomer
in the zone. Combination of zone height and width determinations thus
permits the absolute resolution of the mixed zones into 1-naphthyl-
and 2-naphthylglucuronide concentrations.

We have derived expressions for the calculation of the concentrations of 1-naphthylglucuronide (c_1) and 2-naphthylglucuronide (c_2) (in mmol/l) in solutions injected into the Tachophor

$$c_1 = \frac{2 \cdot (H_2 - H_x) \cdot L_x}{m \cdot (H_2 - H_1) \cdot V} \quad \text{and} \quad c_2 = \frac{2 \cdot (H_x - H_1) \cdot L_x}{m \cdot (H_2 - H_1) \cdot V}$$

where H_1 and H_2 are the relative UV absorbances of the pure 1-naphthyl- and 2-naphthylglucuronide zones, respectively, and H_x is the relative absorbance of the mixed zone. L_x is the zone width of the mixture, and V is the volume of sample injected, in seconds and microlitres, respectively; m is the slope of the calibration plot shown in Fig. 4. It must be stressed that this rather novel method for the resolution of a mixed zone is only possible under conditions of identical zone width: nmole sample injected for the two components, and where the two components possess reasonably different molar extinction coefficients.

In conclusion, the application of analytical isotachophoresis to the assay of enzymatic processes involving the nucleotide UDP-glucuronic acid has been developed one step further by its general applicability to glucuronidation reactions with phenolic acceptors. In most cases, a single isotachopherogram is sufficient to assess the amount of each β-glucuronide formed, and the simultaneous glucuronidation of mixtures of several acceptors can be monitored by this reliable analytical technique. Difficulties are encountered only in special cases, where two β-glucuronides possess virtually identical net mobilities, as with the isomers of naphthol. Nonetheless, the resolution of mixed zones is still possible in such cases.

A prediction which is worthwhile at this early stage of work in this field is that analytical capillary isotachophoresis will find a place in laboratories deeply involved in the study of enzymatic processes where nucleotides play a role as substrates. This technique can probably never replace the much more rapid and less-demanding routine methods such as photometry for enzyme kinetics, but will certainly fill a most important gap for the elucidation of more detailed information on biochemical processes which cannot be gained by other means.

ACKNOWLEDGEMENTS

This work was supported, in part, by the Deutsche Forschungsgemeinschaft. The invaluble help given by LKB-Produkter AB, Bromma, Sweden, and particularly by Astor Baldesten, is gratefully acknowledged. The general advice and encouragement offered by Professors I. Trautschold, and E. and F. W. Schmidt is also gratefully acknowledged.

A. Adam and C. Schots (Editors), *Biochemical and Biological Applications of Isotachophoresis*   187
© 1980 Elsevier Scientific Publishing Company, Amsterdam — Printed in The Netherlands

ISOTACHOPHORESIS OF BRAIN CELL SOLUBLE PROTEIN FROM LEARNING RATS

Paul W. Lange

Insiitute of Neurobiology, University of Göteborg, Fack, S-400
33 Göteborg (Sweden)

INTRODUCTION

Changes in brain protein patterns occur as a result of the establishment of a new behaviour in animals. Several brain regions are involved
in these complicated processes, among which the hippocampus is of out-
standing importance for learning and memory formation. Both membrane-
bound insoluble proteins and soluble proteins have been studied (refs.
1-6). The aim of this investigation was to study changes in soluble
proteins from the hippocampus after a very short training time. As a
behavioural test the transfer of handedness was used. Within minutes
of the start of training, synthesis of brain specific soluble proteins
S 100 and 14-3-2 begins in the hippocampus (refs. 7-9).

In this work isotachophoretic separation of soluble proteins showed
that yet another protein fraction, so far unidentified, increased in
amount as a result of a short training. Addition of Ampholine as spacer
showed that this fraction was homogeneous, at least under the experimental conditions used. It was also found that glycyl-glycine and
asparagine could be used as spacers to separate the protein fractions
from the rest of the soluble protein fractions for a subsequent isotachophoretic preparative separation.

In a separate study it was shown that the amount of the tracer
element zinc in some brain regions changed as a result of a change in
behaviour. The application of isotachophoretic protein separation methods in connection with behaviour and zinc is briefly discussed.

TRAINING METHOD

In this type of behavioural experiment, we used rats which were
induced to transfer handedness in the retrieval of food according to
a previously described test (ref. 10). This consists of a difficult

instrumental learning task in which the rat has to introduce the non-preferred forelimb into a narrow glass tube to retrieve food pills, one by one, by the paw. No surgery or tying of limbs to reverse handedness was used. The original test arrangement of Peterson (ref. 11) was modified in the following way. First, in 23 out of 25 free-choice reaches, the rats showed whether they were left- or right-handed in reaching into a narrow, downward-tilted glass tube to retrieve, one by one, 4 mm diameter protein pills with their preferred paw. The glass tube was kept exactly half full of protein pills by a special filling arrangement. To induce the rats to reverse their handedness, a glass wall was arranged parallel to the glass tube on the side opposite to that of the preferred paw. Space then prohibited the use of the latter, and the rats soon began to use the non-preferred paw. A successful reach is one such that the rat quickly grasps a protein pill by the paw and retrieves it. Control rats were of the same age and weight and were allowed to take the same amount of food by their preferred paw (active controls). The performance curve is linear up to the 7-8th day. The short training in this work consisted of two training periods, each of 25 min, during one day.

MATERIAL

Sprague-Dawley rats weighing 150-175 g were used (13 experimental animals and 8 active control animals). They were killed by decapitation immediately after the second training period. Part of the hippocampus (roughly corresponding to the whole CA3 region) was dissected out using stainless steel microtools. Two samples were taken bilaterally from each brain.

The samples were homogenized in 1.0 mM tris-HCl buffer, pH 7.2, containing 0.03 M NaCl and 0.05 M sucrose (1.0 ml for each hippocampus). The homogenates were centrifuged for 120 min at 30,000 g (0°C). The supernatants were used for isotachophoretic analysis. The protein content was determined by the method of Lowry et al. (ref. 12).

Analytical isotachophoresis in free solution

The LKB 2127 Tachophor equipment was used with ultraviolet (254 nm) and thermal detectors. The amount of protein analysed was about 10 µg in free solution in PTFE capillaries, 0.5 mm I.D. and 23 cm long. The leading solution was 0.005 M HCl, 0.02 M 2-amino-2-methyl-1.3 propandiol, 0.4 % hydroxypropyl-methylcellulose 15,000 cps, pH 9.2.

The terminating solution consisted of 0.005 M ε-amino-n-capron acid, Ba(OH)$_2$, pH 10.9.

RESULTS

We have compared the separation pattern of soluble proteins from the trained rats (2 x 25 min) with that from active controls. The patterns are identical, i.e. no new fractions were formed during the short training (Fig. 1). The central part of the pattern has been divided into four parts, A, B, C and D. Three of them, namely A, C and D, are composed of several fractions whereas fraction B consists of one single peak. This does not necessarily imply that B should be

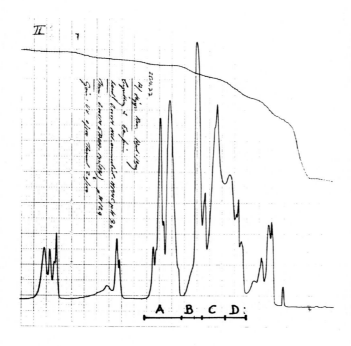

Fig. 1. Analysis in the Tachophor, 23 cm capillary, of 10 μg of soluble protein from the hippocampus of rat. See text.

a homogeneous protein fraction, but only that under the experimental conditions used the resolving power of the isotachophoretic separation is insufficient for further separation. By the addition of various amounts of an Ampholine spacer (Ampholine, pH 3.5-10), however it can be shown that fraction B seems to contain only a single fraction, or, at least, very few protein fractions (the fraction pattern is not shown here). From an isotachophoretic point of view it cannot be excluded that fraction B contains fractions with identical mobilities.

We have also compared the relative amounts of A, B, C and D for trained and control animals (Table 1). The amount of sample (soluble proteins) investigated (measured according to Lowry's method) is in each case 10 μg. This is also seen from the measured surface below A-D, which is the same for trained and control animals. The distribution between the amounts of the fractions A to D, on the other hand, has changed as a result of the short training period. The relative distribution between fractions A, C and D is identical for trained and control animals, as is clearly seen from the values of the ratios A:C:D and their S.E.M. values. The essential result is that the amount of fraction B is significantly increased (P<0.005) in the trained animals in comparison with the controls as a result of the short training.

TABLE 1

Total surface (arbitrary units) below the protein fractions A, B, c and D (see Fig. 1) from the hippocampus of 13 trained rats (Rev = "reversal of handedness") and 8 active controls (PP = "preferred paw"), the percentage distribution of A, B, C and D, and the relative distribution A:B:C:D. All values are given with the S.E.M.

	A+B+C+D	A%	B%	C%	D%	A : B : C : D
Rev.	227±16	31.9±0.90	30.1±1.63	28.5±1.70	9.5±0.33	1.00:0.94±0.058:0.89±0.059:0.30±0.013
PP	242±19	35.1±1.25	24.0±1.15	29.8±0.92	11.1±0.82	1.00:0.68±0.041:0.85±0.040:0.32±0.026

It will be of great interest to characterise fraction B. For that purpose we will perform a preparative isotachophoretic separation of fraction B with the aid of the 7900 Uniphor equipment.

As is discussed elsewhere in detail (see ref. 13), it would be of advantage to find suitable amino acids as discrete spacers for the preparative elution of the B protein fraction.

We have made a series of analytical isotachophoretic separations of brain cell soluble proteins from the hippocampus with additions of various amino acids (leading and terminating electrolytes as given above). Our aim was to find one amino acid with mobility slightly greater than that of fraction B, and one moving somewhat slower. We found the combination glycyl-glycine - asparagine suited our need perfectly (see Fig. 2) Thus we have found the suitable experimental conditions with respect to leading and terminating electrolytes as well as discrete amino acid spacers for the preparative separation of fraction B.

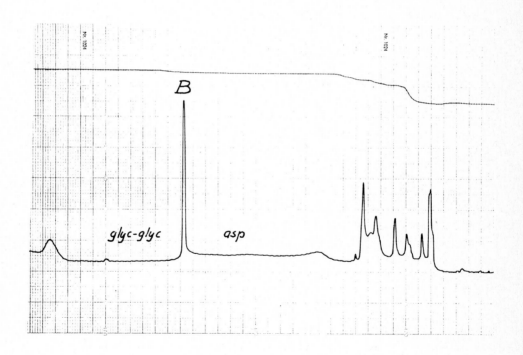

Fig. 2. Analysis in the Tachophor, 23 cm capillary, of 10 μg of soluble protein from the hippocampus of rat. The amino acids glycyl-glycine and asparagine were used as discrete spacers. See text.

DISCUSSION

It is interesting and important to observe that the short training in the initial phase of the establishment of a new behaviour affects only a small part of the soluble proteins. Since no determination of the amount of soluble proteins per dry weight of the hippocampal samples was carried out, it cannot be excluded that all fractions of soluble proteins were synthesized to some extent during the short training period. But is is obvious that fraction B has a unique position with respect to the synthesis of soluble proteins during the short training period. This conclusion is strengthened by the fact that the change in soluble protein pattern takes place in the hippocampus, a brain structure that has no immediate connection with the change of paw or the motor movements of the paws to grasp the food pills in the glass tube as described above.

We intend to use the 7900 Uniphor equipment for a preparative separation of the stimulated B fraction in order to characterize it further. For instance, the possible occurence of subunits can be investigated by SDS-gel electrophoresis, and sequence analysis of the fraction could be of interest.

In connection with this study of the stimulation of synthesis of proteins as a function of a new behaviour, it would be relevant to discuss briefly another approach to this problem.

The mammalian hippocampus contains more zinc than most other regions of the brain. Since the hippocampus is of central importance for learning, it would be of interest to investigate the changes in zinc content in this and other brain regions as s function of learning. In this connection it may be mentioned that adrenal steroids, such as corticosterone, interact with the hippocampus in the presence of zinc (ref. 14).

Normal and adrenalectomized male rats as well as those subjected to a reversal learning test ("transfer of handedness") were therefore investigated with respect to the zinc content in different brain regions. The zinc content was determined by neutron activation analysis.

In normal rats the zinc content, somewhat unexpectedly, did not change in the hippocampus, as a result of four days of training, whereas

it increased markedly in the cortex enthorhinalis. In adrenalectomized
rats, the zinc content both in the hippocampus and in the cortex
entorhinalis changed after training (Table 2). It is highly probable
that the changes in zinc content in the hippocampus and cortex entorhinalis reflect changes in the amount of Zn-enzymes. We intend to
study soluble Zn-enzymes (carboanhydrase, alkali phosphatase and
dehydrogenases) in these brain areas as a function of the establishment
of a new behaviour. To the homogenate of soluble proteins from the
hippocampus and entorhinalis are added the Zn-enzymes, one at the time.

TABLE 2

The amount of Zn$\pm$S.E.M. in the brain from rat (ppm, dry weight)
n = number of animals;_____ = significant difference

Brain region	Normal rats				Adrenalectomized rats			
	Control	n	Trained	n	Control	n	Trained	n
1. Sen-som.cort.	100.0$\pm$3.4	6	100.3$\pm$5.0	3	98.0$\pm$1.9	7	96.0$\pm$2.9	3
2. Vis.cortex	97.8$\pm$2.9	8	95.4$\pm$2.0	5	98.3$\pm$3.5	6	95.7$\pm$2.0	3
3. Hippocampus	102.1$\pm$2.1	14	98.3$\pm$1.9	4	122.5$\pm$3.5	8	85.0$\pm$6.4	2
4. L.coeruleus	54.6$\pm$3.7	8	41.8$\pm$7.1	4	48.3$\pm$7.6	8	52.0$\pm$6.4	3
5. N.ruber	48.7$\pm$3.0	9	48.3$\pm$3.4	3	54.0$\pm$5.1	6	57.5$\pm$3.5	2
6. Subst.nigra	53.3$\pm$6.3	9	63.8$\pm$3.7	5	48.8$\pm$3.0	6	49.0$\pm$6.0	2
7. Cort.enthorh.	106.4$\pm$1.7	5	135.7$\pm$0.9	3	87.0$\pm$4.4	5	104.0$\pm$1.0	2
8. Corp.Call.	59.3$\pm$2.9	9	64.6$\pm$0.9	3	51.0$\pm$4.5	6	51.0$\pm$2.0	2

The homogenate plus the enzyme is fractionated isotachophoretically
(Tachophor 2127), and the position of the Zn-enzymes in the protein
pattern is determined. The amount of Zn in the relevant fractions can
be determined either after a macro-preparative separation (7900 Uniphor)
or a preparative capillary separation (Tachofrac). In the latter case
the Zn content in the enzyme fraction on the acetate strip can be
determined by neutron activation analysis. The Zn-enzymes can also be
determined by the zymogram technique (ref. 15). In all cases, of course,
the results from experimental animals are compared with those from
control animals.

ACKNOWLEDGEMENT

This work was supported by grants from the Swedish Medical Research
Council (N° B78-12X-00086-14C).

194

REFERENCES

1 H. Hydén and P.W. Lange, Science, 159(1968)1370.
2 H. Hydén and P.W. Lange, Brain Res. 22(1970)423.
2 N. Popov, W. Pohle, H.-L. Rütrich, S. Schulzeck and H. Matthies, Brain Res. 101(1975)283.
4 N. Popov, H.-L. Rütrich, W. Pohle, S. Schulzeck and H. Matthies, Brain Res. 101(1976)295.
5 V.E. Shashoua, Brain Res. 111(1976)347.
6 H. Hydén, P.W. Lange and C.L. Perrin, Brain Res., 119(1977)427.
7 H. Hydén and P.W. Lange, Proc. Nat. Acad. Sci. U.S., 65(1970)898.
8 H. Hydén and P.W. Lange, Proc. Nat. Acad. Sci. U.S., 67(1970)1959.
9 K.G. Haglid, H. Hydén and P.W. Lange, J. Neurosci. Res., 1(1975)31.
10 H. Hydén and E. Egyhazi, Proc. Nat. Acad. Sci. U.S., 52(1964)1030.
11 G.M. Peterson, Comp. Psychol. Monogr., 9(1934)1.
12 O.H. Lowry, N.J. Rosebrough, A.L. Farr and R.J. Randall, J. Biol. Chem., 193(1951)265.
13 A. Kopwillem, W.G. Merriman, R.M. Cuddeback, A.J.K. Smolka and M. Bier, J. Chromatogr. 118(1976)35.
14 B. Mc Ewen, personal communication
15 L. Arlinger, J. Chromatogr. 119(1976)9.

A. Adam and C. Schots (Editors), *Biochemical and Biological Applications of Isotachophoresis*   195
© 1980 Elsevier Scientific Publishing Company, Amsterdam — Printed in The Netherlands

ISOTACHOPHORESIS OF THE PROTEINS FROM BOVINE AQUEOUS HUMOR AND
CEREBROSPINAL FLUID, COMPARED TO BOVINE SERUM

J. Bours
Division of Biochemistry of the Eye, Institute for Experimental
Ophthalmology, University of Bonn, D-5300 Bonn-Venusberg (G.F.R.)

P. Delmotte
Neurochemistry Laboratory, National Centre of Multiple Sclerosis,
B-1910 Melsbroek (Belgium)

and G.J. Binkhorst
Clinic for Large Animal Medicine, Yalelaan 16, State University
Utrecht, NL-3584 CM Utrecht (The Netherlands)

SUMMARY

The protein composition of bovine aqueous humor (AH) and cerebrospinal fluid (CSF) and serum was studied by capillary isotachophoresis.
Protein fractions from AH and CSF were quantitated by UV extinction
peak surface integration. Though the overall protein curves from isotachophoresis of AH and CSF showed many similarities, there were significant differences in the percentages of the protein fractions.
Also the Delpech Quotient calculated for AH and CSF was different.

Barrier permeability coefficients from serum to AH and from serum
to CSF were calculated for a number of proteins. From these calculations
it was determined that the permeability coefficients differ between AH
and CSF.

INTRODUCTION

Capillary isotachophoresis, because of the possibility of direct
quantification, represents a major advance in the high-resolution
separation of proteins. This was successfully demonstrated by Delmotte
(ref. 1), who analysed complex protein mixtures from cerebrospinal
fluid (CSF) by this technique. In vertebrate animals two different

ultrafiltration processes from the serum are well known: the formation and production of aqueous humor (AH), and of CSF. To compare these processes it is necessary to study the function of the blood-AH and the blood-CSF barriers regarding the permeability for proteins of different molecular weight.

The aim of the present study was to compare the composition of serum proteins and proteins from AH and CSF, with special regard to the immunoglobulins, for example IgG and IgA. A particular case of interest was the nature and efficacy of the blood-AH and the blood-CSF barriers and whether these barrier-selective filtration processes were of the same nature. Therefore, barrier permeability coefficients from serum to AH and from serum to CSF were calculated for albumin, transferrin, IgA and IgG.

A preliminary report has been given by Bours and Delmotte (ref. 2).

MATERIALS AND METHODS

The age of the animals

Two cows, one aged 5.9 years (N° 1) and the other 7.5 years (N° 2) were available at the veterinary clinic. The age was fixed to known dates of birth.

Collection of serum

Two specimens of blood were collected by venal puncture. Serum was expressed from clotted blood by centrifugation for 10 min at 3000 rpm and 4°C.

Collection and handling of CSF specimens

Two specimens of CSF were collected by suboccipetal puncture from the standing animal, at the point of intersection of the small ligaments and the commissure between the cranial brims of both processes of the atlas. Beforehand, the animal was injected intravenously with a sedative (approx. 1 g of Pentothal, Abbott Spa, Campoverde, Italy) and the spot around the puncture was clean-shaven with spirit-soap and thoroughly cleaned with 70 % ethanol. Superficial and deep anaesthesia was achieved by local injection of 10 ml of lidocaine with 2 % adrenaline. The head of the animal was held by an assistant in a special position, the cranial axis and the neck-band froming an angle of 90°. When standing to the left of the cow's neck a needle invented by Bier (17 cm long and 1.5 mm I.D.) is stuck in at a depth of 10 cm perpendicular to the site of the puncture. The perforation of the dura is experienced as a slight pull of the animal and as a slight resistance which then lessens. After removing the gauching rod of the needle the CSF flows spontaneously, because usually the effective pressure of the CSF is about 170-200 mm. CSF (18-20 ml) was collected with a glass syringe; it was perfectly clear and colourless. There was no evidence of blood contamination. After removal of the needle , the animal's head is immediately stretched. The CSF was frozen and kept at - 20°C.

Collection of aqueous humor.

After CSF collection, four specimens of AH were obtained by aspiration via a limbal puncture (with a N° 20 needle) immediately after enucleation and less than 3 min after the death of the animal. The AH was frozen at -20°C.

Protein measurement

The protein contents of AH and CSF were measured according to the method of Lowry et al. (ref. 3). The AH and CSF were concentrated and simultaneously dialyzed in a Sartorius Membrane Filter (Göttingen G.F.R.). After ultrafiltration the protein content was again measured to determine the concentration factor.

Isotachophoresis

Capillary isotachophoresis was carried out in the LKB 2127 Tachophor, and performed as described by Delmotte (ref. 1). An electronic peak surface integrator was connected to the output of the UV detector to quantitate the protein fractions from AH and CSF.

Permeability coefficients

The permeability coefficients of serum to AH and of serum to CSF were calculated according to:

$$\text{Permeability coefficient} = \frac{\text{integrated area of serum}}{\mu\ell\ \text{serum}} : \frac{\text{integrated area of AH}}{\mu\ell\ \text{of AH x concentration factor}}$$

Delpech quotient

The Delpech quotient (ref. 4) was calculated according to

$$\text{Delpech quotient} = \frac{\text{conc. of IgG from AH}}{\text{conc. of IgG from serum}} : \frac{\text{conc. of albumin from AH}}{\text{conc. of albumin from serum}}$$

RESULTS

The isotachophoretic pattern of serum contains the following major components (Fig. 1c): albumin [1] transferrin [3] IgA + IgG [5] and IgG [6,7] . Three protein preparations from bovine serum indicate as reference the position of the components in the total pattern, as

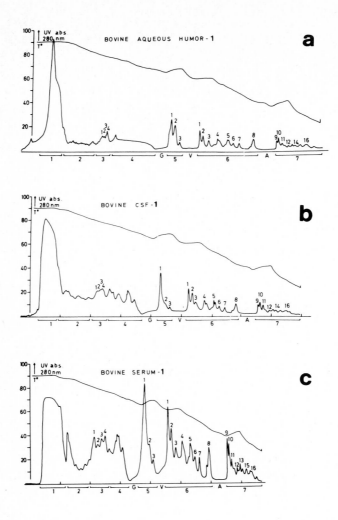

Fig. 1. The patterns of capillary isotachophoretic analysis of (a)
aqueous humor, (AH), (b) cerebrospinal (CSF) and (c) serum
of a 5.9 years old cow. The samples amount to 60 µg of protein.
Fraction 1 = albumin, fraction 2 = serum proteins I, fraction 3
= transferrin, fraction 4 = serum proteins II, fraction 5 =
IgA + IgG, fraction 6 = IgG-I and fraction 7 = IgG-II. The
numerals 1-4, 1-3 and 1-16 indicate the single components
separated: fraction 3, 1-4 = transferrin, fraction 5,1-3 =
IgA + IgG, and fraction 6-7 = IgG. G = glycine, V = valine,
A = β-alanine, T = temperature curve. In (a) fraction 5 from
AH contains only IgG, because is absent in AH.

shown in figure 3 (a) albumin [1],(b) transferrin [3] , and (c) IgG
[6,7] .

   The major part of the protein species in the serum are also present
in the AH and CSF, though at a much lower concentration. The isotachophoretic patterns of AH and of CSF are similar, but nevertheless show
some minor differences (Fig. 1a, b). However, the isotachophoretic
patterns of serum (Fig. 1c), on the hand, and of AH and CSF (Fig 1a, b)
on the other, are quite different both qualitatively and quantitatively.

   The protein contents (in relative percent) of serum, AH and CSF
are shown in Table 1, and are calculated by dividing the peak surface
obtained from the computing electronic integrator by the specific
extinction coefficients (E $_{1\ cm}^{1\%}$) for each protein fraction. Table 2
shows the relative percent distribution of IgG in serum, AH and CSF.
The usual extinction coefficients (ref. 5) were used for the calculations: 1, bovine serum albumin = 6.61 (Fr 1); 2, α_1-antitrypsin,
human = 5.3 (Fr 2); 3, transferrin, porcine = 13.8 (Fr 3); 4, α_1-acid
glycoprotein, human = 8.9 (Fr 4); 5, immunoglobulin A, rabbit = 13.5
(Fr 5 from CSF); 6, immunoglobulin G, human = 13.8 (Fr 5 from AH);
7, immunoglobulin G, human = 13.8 (Fr 6 and 7). Because the E $_{1\ cm}^{1\ \%}$
values of fractions 2-7 are not known for the cow, comparable values
were taken from pig, rabbit and human coefficients (ref. 5).

   The permeability coefficients of the protein fractions 1-7 in figures
1a, b and 2 a,b are calculated according to the formula given in the
methods section, and are given in Tables 1 and 2. The function of the
blood-AH and the blood-CSF barriers can only be compared for their
permeability to proteins of different molecular weight, when the
samples of serum, AH and CSF are taken from the same animal. These
samples are from a 5.9 years' old cow (Fig. 1), and from a 7·5 years'
old cow (Fig. 2) at 20 % and 25 % of their life expectancy (Ref. 6);
respectively.

   Table 2 shows some differences between AH and CSF: the total protein
content for AH is somewhat higher than for CSF. The permeability coefficients of AH are lower than CSF, which indicates the higher total
protein content of AH.

TABLE 1

A comparison of permeability coefficients and protein contents in aqueous humor (AH), cerebrospinal fluid (CSF) and serum between two cows of different ages: (1) 5.9 years' old and (2) 7.5 years' old. IgA = Immunoglobulin A; IgG = Immunoglobulin G, Fr = fraction. Fraction 5 (IgA + IgG) of AH, contains only IgG, due to absence of IgA in AH.

	AH		CSF	
	Cow 1	Cow 2	Cow 1	Cow 2
Total protein (mg/100 ml)	39.00	44.00	34.00	32.00
Concentration factor	11 x	8 x	17 x	18 x
Delpech quotient	1.74	1.39	1.96	2.10

		Permeability coefficient				Protein content (rel (%))					
		AH		CSF		AH		CSF		Serum	
		Cow1	Cow2	Cow1	Cow2	Cow1	Cow2	Cow1	Cow2	Cow1	Cow2
Albumin	(Fr 1)	133	117	165	172	50.2	48.7	43.5	50.1	40.5	43.2
Serum Proteins I	(Fr 2)	164	117	106	150	15.9	17.1	26.4	20.2	15.9	15.1
Transferrin	(Fr 3)	323	241	274	304	3.6	4.1	4.5	4.9	7.0	7.5
Serum Proteins II	(Fr 4)	131	113	144	176	13.3	12.3	13.0	11.9	10.6	10.4
IgA + IgG	(Fr 5)	304	221	493	391	4.4	3.7	2.9	3.2	8.2	6.3
IgG I	(Fr 6)	271	176	350	358	8.2	8.9	6.8	6.6	13.5	11.9
IgG II	(Fr 7)	159	143	260	372	4.4	5.2	2.9	3.1	4.3	5.6
Total	(Fr 1-7)	181	142	204	225	100.0	100.0	100.0	100.0	100.0	100.0
IgG 1-16	(Fr 6-7)	232	164	323	362	12.6	14.1	9.7	9.7	17.8	17.5

The Delpech quotient (ref. 4) is an index for IgG, which is independent of serum IgG concentration. The Delpech quotient differs for AH and CSF (Table 1), which indicates that the barrier permeability from serum to AH and from serum to CSF is different for the two filtration systems.

The permeability coefficients in Table 1 are rather lower for AH than for CSF. This is observed for albumin and IgG (fraction 1, 5, 6, 7) and for other serum proteins (fraction 2, 4) in AH. The subfractions of IgG in Table 2 also show appreciably lower values for AH compared to CSF (subfractions 2, 4-8 of fraction 6 and subfractions 9-16

TABLE 2

A comparison of IgG permeability coefficients and IgG contents in
aqueous humor (AH), cerebrospinal fluid (CSF), and serum between
two cows of different ages: (1) 5.9 years' old and 7.5 years' old.
IgG = Immunoglobulin, Fr = fraction

| | Permeability coefficient | | | | Protein content (rel. %) | | | | | |
| | AH | | CSF | | AH | | CSF | | Serum | |
	Cow 1	Cow 2	Cow 1	Cow 2	Cow 1	Cow 2	Cow 1	Cow 2	Cow 1	Cow 2
IgG 1	538	303	492	540	0.9	1.0	1.0	0.9	2.8	2.3
IgG 2	287	189	381	322	0.9	0.9	0.7	0.8	1.6	1.3
IgG 3	270	378	275	358	0.9	1.1	0.9	0.8	1.4	1.4
IgG 4	245	149	368	258	1.7	1.3	1.2	1.2	2.6	1.5
IgG 5	244	156	326	332	1.3	1.5	1.1	1.0	1.9	1.7
IgG 6	238	158	303	399	0.6	0.7	0.5	0.4	0.9	0.9
IgG 7	181	156	267	371	0.8	0.9	0.6	0.6	0.9	1.2
IgG 8	212	147	326	345	1.1	1.5	0.8	0.9	1.4	1.6
IgG 1-8 (Fr 6)	271	176	350	358	8.2	8.9	6.8	6.6	13.5	11.9
IgG 9	236	239	352	448	0.5	0.5	0.3	0.4	0.7	1.0
IgG 10	202	171	318	433	0.5	0.7	0.4	0.4	0.7	0.9
IgG 11	154	143	243	264	0.7	0.8	0.6	0.7	0.8	0.8
IgG 12	171	116	323	359	0.3	0.4	0.2	0.2	0.4	0.4
IgG 13	149	155	251	389	0.4	0.5	0.3	0.3	0.4	0.6
IgG 14	139	106	320	449	0.5	0.6	0.2	0.2	0.4	0.5
IgG 15	117	120	195	343	0.6	0.7	0.4	0.4	0.4	0.6
IgG 16	101	117	177	370	0.9	1.0	0.5	0.5	0.5	0.8
IgG 9-16 (Fr 7)	159	143	260	372	4.4	5.2	2.9	3.1	4.3	5.6
IgG 1-16 (Fr 6-7)	232	164	323	362	12.6	14.1	9.7	9.7	17.8	17.5

of fraction 7). This corresponds to higher percentages of albumin and
IgG in AH.

Contrary of the CSF, the AH of cows 1 and 2 shows age difference in
Tables 1 and 2. The permeability coefficient for AH decreases in 1.6
year, which may be an indication of increased permeability for IgG;
there may be no appreciable development of IgG within the AH independent
from the serum IgG (ref. 4).

The heterogeneity of IgG in serum, AH and CSF (16 subfractions !)
shows a difference in the distribution of the electrophoretic mobility
of these subfractions.

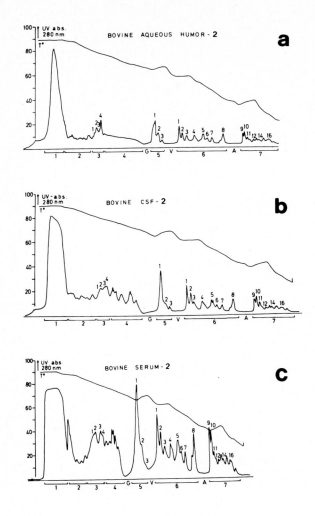

Fig. 2.   The patterns of capillary isotachophoretic analysis of (a)
aqueous humor (AH), (b) cerebrospinal fluid (CSF) and (c)
serum of a 7,5 years' old cow. For further explanations
see the legend to Fig. 1.

DISCUSSION

Generally the isotachophoretic patterns of bovine AH and CSF (Figs.
1a, b and 2a, b) resemble the patterns of human CSF (ref. 1). Some
quantitative differences are, however, reported (ref. 1): in the cow
a lower transferrin content, a higher IgA content in bovine CSF, and
a different distribution in bovine AH and CSF compared to human CSF.

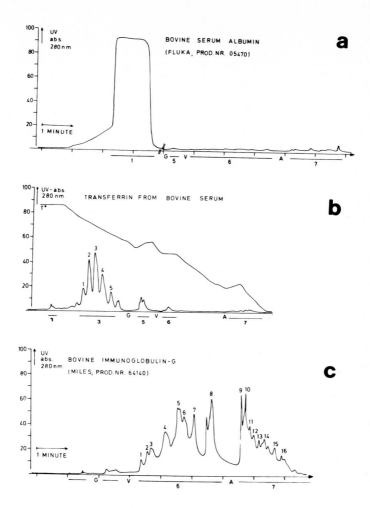

Fig. 3. The patterns of capillary isotachoretic analysis of (a)
bovine serum albumin (20 μg), (b) transferrin from bovine
serum (6 μg) and (c) immunoglobulin G from bovine serum
(20 μg). Fraction 1 = albumin, fraction 3 = transferrin
(purity 83 %), fraction 5 = IgA + IgG, fraction 6 = IgG-I,
fraction 7 = IgG-II. The numerals 1-5 and 1-16 indicate the
single components separated: 1-5 = transferrin, 1-16 = IgG.
G = glycine, V = valine and A = β-alanine.

In 1956 Davson (ref. 7) emphasized the similarity of the blood-ocular and blood-CSF barriers. There are, however, some differences between AH and CSF since the CSF contains all serum proteins albeit in 100-fold lower concentrations (ref. 8), whereas the barrier between serum and CSF is characterized by a true percolation process. The AH does not contain IgA and IgM as long as the blood-AH barrier is intact (refs 9 and 10). According to our present immunologic investigations (ref. 2) no IgA (mol.wt. = 170,000) and no IgM (mol.wt. = 900,000) could be detected by immunoelectrophoresis in concentrated samples of bovine AH by monospecific antisera. Neither have we detected the enzyme aldolase (E.C. 4.1.2.13) (mol.wt. = 158,000) in AH contrary to Hockwin and Kietzman (ref. 11) who found aldolase in appreciable amounts in bovine AH taken at least 10 min after death (ref. 11). IgG (mol.wt. 145,000 to 150,000) still passes the barrier. This shows that the serum-AH barrier has a transfer limit which reaches a molecular weight of 150,000.

CONCLUSIONS

(1) The blood-AH and the blood-CSF barriers have many similarities.
(2) The nature of the selective filtration processes for AH and CSF differ in detail.
(3) In principle, all serum proteins are present in the CSF. The serum-AH barrier in the cow has a transfer limit at a molecular weight of 150,000.

ACKNOWLEDGEMENT

The authors are indebted to Dr. Th. Wensink (Clinic for Large Animal Medicine, Utrecht) for his kind cooperation.

This work was supported by the "Deutsche Forschungsgemeinschaft" (Ho 249/8/12; Oh 32/2).

REFERENCES

1 P. Delmotte, Sci. Tools, 24(1977)33-41.
2 J. Bours and P. Delmotte, Ophthalmologica (Basel), (1979) in press.
3 O.H. Lowry, N.J. Rosebrough, A.L. Farr and R.J. Randall, J. Biol. Chem., 193(1951)265-283.
4 B. Delpech and E. Lichtblau, Clin. Chim. Acta, 37(1972)15-23
5 D.M. Kirschenbaum, in H. Sober (Editor), Handbook of Biochemistry, Chemical Rubber Co., Cleveland Ohio, 2nd ed., 1970
6 O. Hockwin, H. Fink, H. Schallenberg and F. Rast, V. Albrecht van Graefes Arch. Ophthalmol. 195(1975)17-26.
7 H. Davson, Physiology of the Ocular and Cerebrospinal Fluids, Churchill, London, 1956.

8 E.S. Mingioli, W. Strober, W.W. Tourtelotte, J.M. Whitaker and
   D.E. McFarlin, Neurology, 28(1978)991-995.
9 J.P. Dernouchamps and J.F. Heremans, Exp. Eye Res., 21(1975)
   289-297.
10 J.P. Dernouchamps, J.P. Vaerman, J. Michiels and P.L. Masson,
   Amer. J. Ophthalmol. 84(1977)24-31.
11 O. Hockwin and M. Th. Kietzmann, Albrecht von Graefes Arch.
   Ophthalmol., 206(1978)179-181.

A. Adam and C. Schots (Editors), *Biochemical and Biological Applications of Isotachophoresis*   207
© 1980 Elsevier Scientific Publishing Company, Amsterdam — Printed in The Netherlands

THE CRYSTALLIN COMPOSITION OF BOVINE LENS EPITHELIUM, EQUATOR AND
NUCLEUS, IN RELATION TO AGING: A COMPARISON OF ISOTACHOPHORESIS,
ISOELECTRIC FOCUSING AND ANTIGEN/ANTIBODY CROSSED ELECTROPHORESIS

J. Bours
Division of Biochemistry of the Eye, Institute for Experimental
Ophthalmology, University of Bonn, D-5300 Bonn-Venusberg (G.F.R.)

and

H.-D. Zauzig, H. Rink
Institute for Radiobiology, University of Bonn, D-5300 Bonn-Venusberg
(G.F.R.)

with the collaboration of P. Delmotte
Neurochemistry Laboratory, National Centre for Multiple Sclerosis,
Van Heylenstraat 16, B-1910 Melsbroek (Belgium).

ABSTRACT
    Antigen/antibody crossed electrophoresis of bovine lens epithelial
cell extracts reveals one pre-α-crystallin, one or two α-crystallins,
four β-crystallins and one γ-crystallin. With aging of the epithelium
and the nucleus, the α-crystallin precipitin arc shifts position towards
the direction of highest electrophoretic mobility. By isoelectric
focusing several α-crystallin components, twelve groups of β-crystallin
components and eight γ-crystallin components are detected in the epithelium. Significant differences are found both in isoelectric point
and in concentration as shown by the densitometer values of isofocused
single γ-crystallins of the epithelium, equator and nucleus.

    By capillary isotachophoresis of the epithelium at least one α-
crystallin, fourteen β-crystallin components and sixteen γ-crystallin
components are detected. The young lens equator and nucleus show a
similar number of components, but the concentrations in the lens
epithelium, equator and nucleus differ considerably.

With aging of the lens, γ-crystallin concentration decreases and high molecular weight β-crystallin concentration increases in all lens parts including the epithelium. The epithelial γ-crystallins differ structurally from the γ-crystallins in the equator and the nucleus.

INTRODUCTION

Previously we demonstrated by one-dimensional immunoelectrophoresis using various antisera that the crystallins from bovine lens epithelium consist of two α-crystallins, three β-crystallins and one γ-crystallin (ref. 1). In the present study we confirm these data by application of two other electrophoretic methods with higher resolution, i.e. isoelectric focusing and isotachophoresis, and obtain additional useful information. The following subjects were studied:

(1) The number of crystallin precipitin lines of the aging bovine lens epithelium, equator and nucleus as detected by two immuno-electrophoretic methods;

(2) The crystallins from the epithelium, the equator and the nucleus of the bovine lens, in relation to aging, as detected by iso-electric focusing and isotachophoresis;

(3) The nature of the γ-crystallins in bovine lens epithelial cells of three different age groups as detected by isoelectric focu-sing and isotachophoresis.

A preliminary note on the crystallin composition of the bovine lens epithelium was published by Bours et al. (ref. 2).

MATERIALS AND METHODS

The bovine lenses were prepared, their fresh weights determined, and they were separated into three age groups: (a) < 1 year old, (b) 1-10 years'old, and (c) > 10 years'old. Lens capsules, trypsinized lens cells and epithelial cell crystallin extracts were treated as described by Bours et al (ref. 1). All other crystallins from the lens equator and nucleus were prepared according to the description of Bours et al. (ref. 3).

The polyvalent and specific antisera were prepared as published (ref. 1). Two-dimensional antigen/antibody crossed electrophoresis was performed according to the method of Bours and Brahma (ref. 4). Isoelectric focusing was carried out according to the method of Bours

(ref. 5). The epithelial samples used weighed 1 mg, and each contained 400 µg of protein, as determined by the method of Lowry, modified
by Schacterle and Pollack (ref. 6). All other samples contained 300 µg
of protein.

Capillary isotachophoresis of the crystallins was performed with
an LKB 2127 Tachophor, equipped with the standard capillary tube
(23 cm long, 0.45 mm internal diameter, thermostatted at 12°C) (ref. 7).
Quantitative estimation of the crystallin fractions (monitored by UV
absorption at 280 nm) was carried out by peak surface integration with
a Spectra-physics migrator (Model 602). This electronic integrator
was programmed to hold its baseline from the start until the end of
each run and to separate the protein peaks by verticals dropped at
the lowest point between two peaks. The composition of the different
electrode solutions, the composition of the spacer mobility gradient
and all the details of the working procedure are described elsewhere
(ref. 7). The separation pH was 9.1. A 3-µℓ volume of sample solution
(1 % w/v total protein content) was injected together with 2 µℓ of
the spacer mobility gradient solution. The mobility marker amino acids
were glycine, valine and β-alanine. Each run was started with the
power supply in the constant voltage mode at 7 kV, which initially
supplied a current of about 150 µA. When the current dropped to 40 µA,
the power supply was manually switched to the constant current mode,
usually 12 to 13 min after the start of the run. During detection,
under a constant current of 40 µA, the voltage rose from about 8.5 kV
to about 11 kV. Detection occurred between 19 and about 26 min after
sample injection using a recorder chart speed of 5 cm/min.

RESULTS

Antigen/antibody crossed electrophoresis of epithelial antigens
with polyvalent and anti-α-crystallin serum shows one pre-α-crystallin
(Fig. 1a-c), one or two α-crystallins (Fig. 1a-d, h), four β-crystallins
and one γ-crystallin (Fig. 1a-d). In the equator and nucleus of the
youngest lens one more pre-α-crystallin is detected (Fig. 1k,m) whereas
in the aged nucleus the pre-α-crystallins are absent (Fig. 1l). In
the youngest epithelium (<1 year old), the α-crystallin is present
in two forms, α_1 and α_2, which have the same immunological determinants
demonstrated by polyvalent (Fig. 1b,d) and monovalent antiserum (Fig.
1h). In the older lens only one α-crystallin is found in the lens epithelium (Fig. 1a, c), in the equator (Fig. 1j, k, n, o), in the nucleus

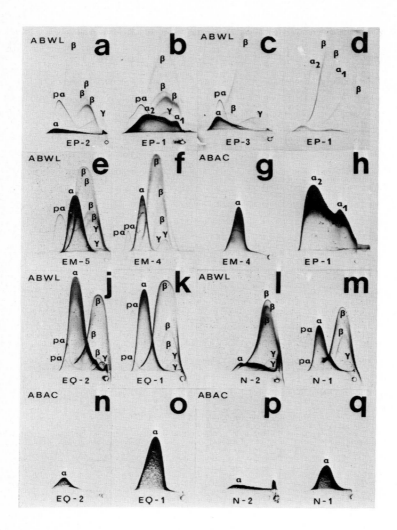

Fig. 1. Antigen/antibody crossed electrophoresis of bovine lens epithelium, equator, nucleus and embryonic lens against polyvalent
and anti-α-crystallin serum. EP-1 = epithelial cells, < 1
year old, EP-2 = epithelial cells, 1-10 years'old; EP-3 =
epithelial cells, > 10 years'old; EM-4 = embryonic lens, 4
months'old; EM-5 = embryonic lens, 4.5 months'old; EQ-1 =
equator, 0.2 years'old; EQ-2 = equator, 25 years'old; N-1 =
nucleus, 0.2 years'old; N-2 = nucleus, 25 years'old. ABWL =
anti-bovine whole (calf) lens serum; ABAC = anti-bovine lens
cortex α-crystallin serum; pα = pre-α-crystallin, α = α-
crystallin, β = β-crystallin, γ = γ-crystallin.

(Fig. 1l, m, p, q) and in the embryonic lens (Fig. 1e-g). There is also an aging phenomenon in the epithelium (Fig. 1a, c) and the nucleus (Fig. 1l, p), which affects the electrophoretic mobility of α-crystallin: with increased age, the mobility shifts markedly towards the direction of the anode.

All four β-crystallins that are present in the epithelium (Fig. 1a-d) are also found in the other lens parts (Fig. 1j-m). The γ-crystallins seem to be present in only one form in the epithelium (Fig. 1a-d), whereas the nucleus, the young equator and the embryonic lens have two immunologically detectable forms of γ-crystallin (Fig. 1e, f, k-m) (4,8). The number of precipitin lines for each crystallin detected in the epithelium, equator and nucleus, is arranged conveniently in Table 1 [where refs. 9-12 are referred to].

TABLE 1
The number of crystallin components of the aging bovine lens in epithelium, equator and nucleus as detected by immunoelectrophoresis, by antigen/antibody crossed electrophoresis, by isoelectric focusing and by isotachophoresis.
EP = Epithelium, EQ = Equator, EM = Embryonic lens, N = Nucleus,
P = Polyvalent, M = Monovalent, T = Tetravalent, - = not detected,
+ = positive reaction

Separation technique	Protein species	Tissue: Age (years)	EP 1	EP 1-10	EP 10	EQ 0.2	N 0.2	EQ 25	N 25	EM 0.4	Antisera: P	M	T	Reference
2D AG/AB Crossed electrophoresis	pre-α-crystallin 1		1	1	1	1	1	1	0	1	+			1,4,8
	pre-α-crystallin 2		-	-	-	1	1	-	-	1	+			4,8
	α-crystallin		2	1	1	1	1	1	1	1	+	+		1,4,8
	β-crystallins		4	4	4	4	4	4	4	4	+		+	1,4,8
	Low Mol. Weight β-crystallins		1	1	1	1	1	1	1	1		+		9
	γ-crystallins		1	1	1	2	2	1	2	2	+			1,4,8
	embryonic γ-crystallins		1	1	1	1	1	1	1	1			+	1,4,8
Isoelectric focysing	α-crystallin components		12	12	13	9	8	11	9					3,5,8,10,13
	pre-α-crystallin		1-2	1-2	1-1	1-2	1-2	0	0					3,5,8,10,13
	β-crystallin groups		12	11	10	12	10	10	12					3,5,8,10,13
	β-crystallin components		20	21	22	18	17	17	21					3,5,8,10,13
	γ-crystalllisn		8	5	3	5	8	0	4					3,5,8,10,13
Isotacho-phoresis	pre-α-crystallin		1	1	1	1	1	0	0					7,11,12
	α-crystallins		1	1	1	2	2	2	2					7,11,12
	β-crystallins		14	13	13	12	14	12	13					7,11,12
	γ-crysrallins		16	15	15	15	14	15	15					7,11,12

Isoelectric focusing of the epithelial crystallins shows the following features (Fig. 2). α-Crystallin is more concentrated in the younger than in the older epithelium (Fig. 2c-e). The number of β-crystallins is more or less comparable in the epithelium of the three age groups (Table 1). The oldest epithelium (Fig. 2e) shows the highest concentration of β_{1-6}-crystallins which is an aging effect. The γ-crystallins in the epithelium differ structurally, qualitatively and quantitatively from the γ-crystallins present in the equator and in the nucleus. Isoelectric focusing shows striking differences in appearance and concentration of the various γ-crystallin components in each lens part (Fig. 2); their densitometric values are given in Table 2. Additionally, concomitant with aging one sees a continuous decrease of the γ-crystallin concentration in the nucleus, the equator and also in the epithelium (Table 2). In the epithelium the decrease of γ-crystallins is only moderate compared to this effect in the nucleus, while in the equator the γ-crystallins disappear almost completely. This is visualized in Fig. 3 which gives the densitometer integrator units from each lens part, from animals of different ages.

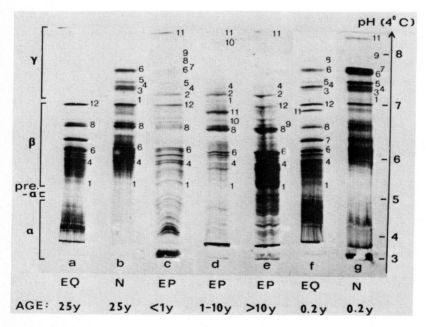

Fig. 2. Thin-layer isoelectric focusing of epithelial cells of three age groups compared to the equator and nucleus of the bovine lens. EP = epithelium, EQ = equator N = nucleus, α = α-crystallins, pre-α = pre-α-crystallin, β = β-crystallins, γ = γ-crystallins.

Table 2

The densitometer values of eleven γ-crystallin components after thin-layer isoelectric focusing of bovine lens epithelial cells of three different age groups, compared to young and old lens equator and nucleus

Lens part	Age (years-	Densitometer values γ-Crystallin component number											Σ of γ-crystallin 1-11
		1	2	3	4	5	6	7	8	9	10	11	
Nucleus	0.2	5	0	46	56	59	70	70	0	0	0	15	321
Nucleus	25	10	0	28	28	16	38	0	0	0	0	0	120
Equator	0.2	0	0	20	32	10	42	0	11	0	0	4	119
Equator	25	0	0	0	0	0	0	0	0	0	0	0	0
Epithelium	<1	9	22	0	0	14	6	7	3	6	0	20	87
Epithelium	1-10	10	29	0	19	0	0	0	0	0	3	3	64
Epithelium	>10	0	26	0	18	0	0	0	0	0	0	5	49

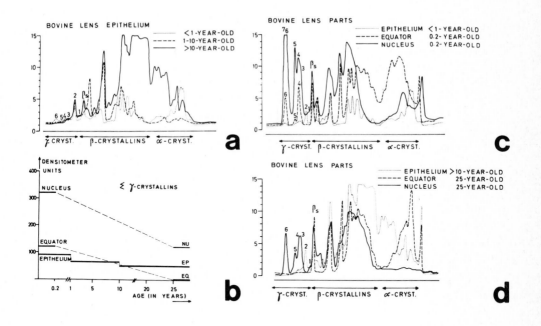

Fig. 3. Densitometric evaluations of isoelectric focusing patterns from epithelium, equator and nucleus of the aging bovine lens. 0-15 = recorder scale; 100-400 = the number of densitometer units read from the integrator.

When the densitometric scans of the isofocused epithelial samples from three age groups are superimposed (Fig. 3a), the differences between the three curves are easily observed. The γ-crystallin components N° 2 and 9 are absent in the equator and nucleus (Fig. 2b, f, g; 3a), whereas γ-crystallin component N° 3 is absent in the epithelium (Fig. 2c-e; 3a, c; Table 2).

Another important aspect of aging is the considerable increase of the β_{1-6}-crystallin concentration in the epithelium after age 10. This is shown in Fig. 3a where β_8-crystallin concentration is also shown to be higher than in the two younger epithelial preparations. In Fig. 3c the three lens parts from animals of approximately the same age are compared. The young nucleus shows the highest concentration of γ-crystallin. This concentration is only moderate in the equator and of different composition than the nucleus: the γ_1- and γ_7-crystallin components seem to be absent in the equator (Fig. 2f; 3c). The β-crystallins are more concentrated in the young nucleus, while the α-crystallins are more prominent in the young equator (Fig. 3c). The aged lens in Fig. 3d shows a decrease of the γ-crystallin concentration in all lens parts (Fig. 3b, c) and an increase of the β_{1-6}-crystallin concentration. In the aged nucleus, the α-crystallins of lower molecular weight are almost absent because with aging they are increasingly transferred to the α-crystallins of higher molecular weight (14) which cannot be visualized by the thin-layer isoelectric focusing technique (10).

The isotachophoretic experiments (Figs. 4 and 5) show a somewhat higher resolution than isoelectric focusing of the β- and γ-crystallin component. The order of electrophoretic mobility of the proteins is generally the same in isotachophoresis and isoelectric focusing. An exception is the pre-α-crystallin which appears ahead of the α-crystallins in isotachophoresis. The first γ-crystallin appears after the amino acid spacer valine. An essential marker in this respect is the β_s-crystallin, which appears between the amino acid spacers glycine and valine (Fig. 4a-d). The β-crystallins of higher molecular weight are positioned ahead of the amino acid spacer glycine. The conclusions drawn from isoelectric focusing are confirmed by the isotachophoretic experiments.

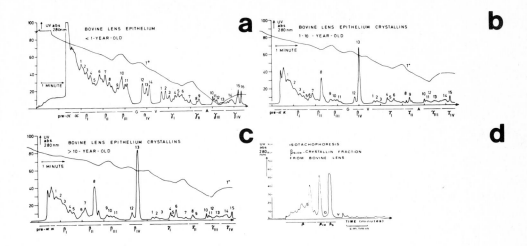

Fig. 4. The patterns of capillary isotachophoretic analysis of bovine
lens epithelium: (a) < 1 year old; (B) 1-10 years'old; (c)
> 10 years'old. These samples amount to 60 µg of protein.
β_{I-IV} and γ_{I-IV} denote the groups of β- and γ-crystallin
components. The arabic numerals 1-14 and 1-16 indicate the
separated individual crystallin components; (d) β_s-fraction
from bovine lens: $\beta_s = \beta_{slow}$-crystallin; β_{LM} = low-molecular-
weight β-crystallin. pre-α = pre-α-crystallin, α = α-crystal-
lins, β = β-crystallins, γ = γ-crystallins. G = glycine,
V = valine, A = β-alanine.

The amounts of the individual water-soluble crystallins have been
calculated as per cent water-soluble (% WS, Fig. 6). The pre-α-
crystallin is prominent in the epithelium (Figs. 4a-c and 6), but
has a low concentration in the equator and nucleus of the young lens,
and is absent in the aged equator and nucleus. With aging the α-crystallins decrease in the epithelium, increase in the equator and decrease in the nucleus. The β-crystallins, which have about the same
concentration in the three lens parts, increase considerably in the
epithelium, equator and also in the nucleus (Fig. 6). The increase
of the β_{13}-component in the epithelium with aging is striking (Fig.
4a-c). The computed integrator values of the epithelial total γ-crys-

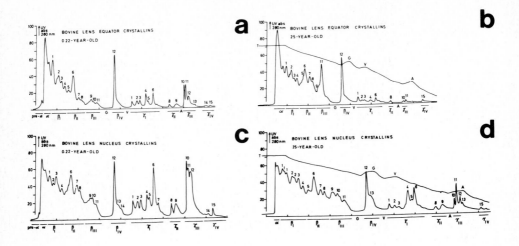

Fig. 5. The patterns of capillary isotachophoretic analysis of the
crystallins from the bovine lens equator and nucleus; (a)
equator, 0.22 years'old, (b) equator, 25 years'old, (c)
nucleus, 0.22 years'old, (d) nucleus, 25 years'old. I-IV;
1-12 and 1-15 are explained in the legend to Fig. 4.

tallins show a decrease which is comparable to the decrease of the
densitometer values after isoelectric focusing (Fig. 3b). When recalculated as % WS, the γ-crystallins in the epithelium of young and
aged lens show about the same concentration, but in the equator and,
to a lesser extent in the nucleus, the reduction in concentration
of the crystallins is considerable (Fig. 6).

It should be emphasized, however, that the percentages given in
Fig. 6 are only relative, due to the increasing concentration of the
albuminoid with advancing years (cf. refs. 11 and 12), which is not
taken into account in this figure. When the absolute albuminoid concentration is considered, the β-crystallins actually show a slight
decrease with age, rather than increasing as they appear to do in

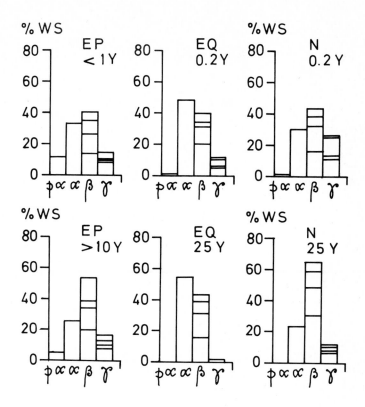

Fig. 6. Distribution of α-, β- and γ-crystallins and their subfractions
in bovine lens epithelium, equator and nucleus during aging,
evaluated by a computing integrator system of the isotachophoretic patterns from Figs. 4 and 5. The divisions in the columns
representing the β- and γ-fractions indicate the relative a-
mounts of each subfraction that were obtained. These subfractions
are, from bottom to top, I - IV for the β- and for γ-crystallins
Further explanations are given in the legends of Figs. 4 and
5. The computed peak surfaces obtained by isotachophoresis were
divided by the specific extinction coefficients ($E_{1cm}^{1\%}$)of the
bovine lens (18): for pre-α-crystallin = 22.33, α-crystallins =
8.30, $β_H$-crystallins 24.75, $β_L$-crystallins = 22.33, γ-crystallins = 22.64 % WS = percentage of water-soluble crystallins,
EP = epithelium, EQ = equator, N = nucleus.

figure 6. The γ-crystallins decrease whether one considers the albuminoid or not (refs. 11-13).

DISCUSSION

Bloemendal and Vermorken (ref. 15) have demonstrated that the α-crystallin composition of the epithelium is not homogeneous, rather it differs from the central to the pre-elongation zones. We did not discriminate between these areas of the epithelium because we isolated the whole capsule including the epithelial cells of the pre-elongation zone. However, we determined microscopically that our epithelial preparations were free from fibre cells (refs. 1 and 16).

In lens cell differentiation, the epithelial cells transform into the fibre cells of the lens cortex. Papaconstantinou (ref. 17) found that this transition is accompanied by a qualitative change in the synthesizing ability of proteins and postulated that γ-crystallin is the biochemical marker of differentiation. This hypothesis is now questionable, because we have detected γ-crystallins in the epithelium (Figs. 1-6) (ref. 1). It must be noted, however, that the epithelial γ-crystallin components (Figs. 2c-e, and 4a-c) differ structurally from the γ-crystallins in the equator (Figs. 2a, f; 3c, d and 5a, c) and the nucleus (Figs. 2b, g; 3c, d and 5b, d). Furthermore, γ-crystallins present in the aged epithelium (Figs. 2e and 4c) are completely lost during differentiation in aged equator (Figs. 2a and 5b).

Isotachophoresis is comparable to isoelectric focusing in resolving power, but has the considerable advantage of direct quantification of the protein fractions by integration of the UV 280 nm absorption peaks. The over-all reproducibility of the results of isotachophoresis is demonstrated for the bovine lens in this article (Figs. 4 and 5), and for the mouse lens by Delmotte (ref. 7) and Bours and Delmotte (ref. 12)

We did confirm the observations of Iwig and Glaesser (ref. 19) who found that α-crystallin in the young epithelial cells consists of two entities (Fig. 1b, d, h) which differ in electrophoretic mobility and also in diffusion rate (ref. 1), the latter being related to the molecular weight of the protein.

During aging of the lens three different and continuous processes can be followed with exactitude:

(1) The decrease in γ-crystallin concentration (Figs. 3b, 4 and 5) and loss of certain γ-crystallin components in all lens parts (Figs. 2, 4 and 5) (refs. 3 and 10);

(2) The increase of the β-crystallin components (β_{1-6}) of higher molecular weight in all lens parts (Fig. 6) (refs. 3 and 10);

(3) The increase of anodic mobility of α-crystallin form the epithelium and from the nucleus. This increase is most probably associated with the continuous increase in molecular weight of the α-crystallin entities (Fig. 1) (ref. 14).

CONCLUSIONS

(1) The epithelial cells contain one pre-α-crystallin, two α-crystallins, four β-crystallins and one γ-crystallin as detected immunologically;

(2) The epithelial α-crystallin of the young lens shows a different behaviour in electrophoresis and consists of two entities with identical determinants but different diffusion rates. The epithelial α-crystallin of the aging lens is subject to the same aging processes that take place in other lens parts;

(3) The β-crystallins from the lens epithelium show a similar behaviour to that of the β-crystallins in the other lens parts: a significant increase of the high-molecular-weight β-crystallins is observed with aging;

(4) γ-Crystallin is detectable in the lens epithelium, but is different from the γ-crystallins detected in the equator and nucleus;

(5) With aging the γ-crystallin content decreases only slightly in the epithelium but significantly in the equator and also, though to a lesser extent, in the nucleus.

ACKNOWLEDGEMENTS

The authors are grateful to Mrs. Elke Oellers and Miss Barbara Polenz for the photographic illustrations. They are also very grateful to Miss Dr J. K. Jose (University of California, Berkeley) for her helpful suggestions.

This study was supported by the "Deutsche Forschungsgemeinschaft" (Oh 32/i, Ri 285/3).

REFERENCES

1 J. Bours, H.-D. Zauzig and H. Rink, Ophthal. Res., (1979) (in press).
2 J. Bours, H.-D. Zauzig and H. Rink, Ophthal. Res., 10(1978)319-320. Presented at the AER Congr. Amsterdam, 1978.
3 J. Bours, K. Doepfmer and O. Hockwin, Docum. Ophthal. Proc. Ser., 8(1976)75-89.

4 J. Bours and S.K. Brahma, Exp. Eye Res., 16(1973)131-142.
5 J. Bours, In B.J. Radola and D. Graesslin (Editors) Electrofocusing and Isotachophoresis. Walter de Gruyter, Berlin, New York, 1977, pp. 303-312.
6 G.R. Schacterle and R.L. Pollack, Anal. Biochem., 51(1973)654-655.
7 P. Delmotte, Sc. Tools, 24(1977)33-41.
8 J. Bours, In O. Hockwin and H.-R. Koch (Editors) Arzneimittelneben-wirkungen am Auge. Fischer Verlag, Stuttgart, 1977, pp. 270-280.
9 J. Bours and M. Rabaey, Exp. Eye Res., 20(1975)180-181.
10 J. Bours, A. Weber and O. Hockwin, In Y. Courtois and F. Regnault (Editors) INSERM Symp. Biology of the epithelial lens cells in relation to development, aging and cataract. INSERM, Paris 60 (1976)205-213.
11 J. Bours, Interdiscipl. Topics Gerontol., 12(1978)196-204.
12 J. Bours and P. Delmotte, Sc. Tools, 1980) in press.
13 J. Bours Comp. Biochem. Physiol. B., (1979) in press.
14 J. Bours, A. Wieck and O. Hockwin, Interdiscipl. Topics Gerontol., 12(1978)205-220.
15 H. Bloemendal and A.J.M. Vermorken, Interdiscipl. Topics Gerontol., 12(1978)41-49.
16 M. Iwig and D. Glaesser, Ophthal. Res., 4(1972/73)328-342.
17 J. Papaconstantinou, Biochem. Biophys. Acta, 107(1965)81-90.
18 G.J. van Kamp, Thesis, University of Nijmegen, 1973, p. 24.
19 M. Iwig and D. Glaesser, Ophthal. Res., 10(1978)67-72.

A. Adam and C. Schots (Editors), *Biochemical and Biological Applications of Isotachophoresis*   221
© 1980 Elsevier Scientific Publishing Company, Amsterdam — Printed in The Netherlands

AGE-DEPENDENT VARIATIONS IN THE COMPOSITION OF THE CRYSTALLINS AND
ALBUMINOID OF THE MOUSE LENS, STUDIED BY ISOELECTRIC FOCUSING AND
ISOTACHOPHORESIS

J. Bours

Division of Biochemistry of the Eye, Institute for Experimental
Ophthalmology, University of Bonn, D-5300 Bonn-Venusberg (G.F.R.)

and

P. Delmotte

Neurochemistry Laboratory, National Centre for Multiple Sclerosis,
Van Heylenstraat 16, B-1910 Melsbroek, Belgium

SUMMARY

   Wet and dry weights of the normal mouse lens, and its composition
of crystallins and albuminoid were studied. Thin-layer isoelectric
focusing and capillary isotachophoresis were carried out to separate
crystallin proteins from the mouse lens into pre-α-crystallin, α-
crystallins, 14 β-crystallin components and 16 γ-crystallin components.
With aging a relative increase and an absolute decrease in the water-
soluble crystallins and a relative and absolute increase in the water-
insoluble albuminoid were observed. The concentrations of the low
molecular weight crystallins, i.e. the high isoelectric point β-
crystallins and the γ-crystallins, were reduced considerably with
aging.

   Isotachophoresis is comparable to isoelectric focusing in resolving
power, but has the added advantages of quantitative sample application
plus direct quantification of the protein fractions by integration of
the UV 280 nm absorption peaks.

INTRODUCTION

   Capillary isotachophoresis is a very reproducible and rapid analytical method for quantification of proteins (ref. 1). Only minute amounts
(about 30 μg) of protein are required. When Ampholine carrier ampholytes

are used to create a spacing mobility gradient, capillary isotachophoresis shows high resolving power for protein separation. Hemoglobins
are also well separated in this way (refs. 2 and 3). We have added
specific amino acids (glycine, valine and β-alanine) to the Ampholine
as discrete spacers, in a method similar to that described by Kjellin
and co-workers (refs. 4 and 5) in the analysis of cerebrospinal fluid.
Kopwillem et al. (ref. 6) and Moberg et al. (ref. 7) also applied amino
acids as individual spacer ions to obtain optimal separation of human
serum proteins. In the present work hydroxypropylmethylcellulose was
also used to improve zone stabilization (refs. 2 and 4).

The aim of the present study is to compare capillary isotachophoresis
(refs. 1 and 8) and isoelectric focusing (ref. 9), since both of these
methods possess high resolution. The resolving power of isoelectric
focusing is based on the isoelectric charge of the well-separated proteins at their focused position, which is also their state of minimum
energy. The resolving power of isotachophoresis is based on the
separation of charged particles (proteins) which initially move at
different speeds so that they become separated according to their
electrophoretic mobilities (ref . 10).

Isotachophoresis has a somewhat higher resolving power than isoelectric focusing and increases the number of separated protein components. Consequently the technique provides new information on the
changes of protein composition of eye lens crystallins with age. A
related study has been reported concerning the effect of aging on the
rat lens crystallins (ref. 11).

In this study, the wet weight of the lens of the mouse Mus musculus
was determined and then treated by a general formula from which it was
possible to determine the age of an animal. This method of determining
age from the wet weight of the lens was published by Hockwin et al.
(ref. 12) who applied the technique to various other animal species.

EXPERIMENTAL
Materials
    Normal lenses from mice of known age were removed from enucleated
eyes immediately after slit-lamp examination and killing the animals
(specified pathogen-free albino mice NMR I, from the Central Institute
for Laboratory Animals, Hannover, G.F.R.). The wet weight of the lenses

was determined immediately, and then the lenses lyophilized and the dry weight determined. Thereafter the dry lenses were homogenized (Ultra-Turrax) in distilled water at 4°C and centrifuged at 38,000 x g for 1 h at 4°C and the resulting supernatant fluids were lyophilized. The sediments were washed repeatedly with distilled water to free them from water-soluble materials. The resulting supernatant fluids and the sediments were lyophilized separately and the water-soluble material (crystallins) and the water-insoluble material (albuminoid) were weighed. The weights are given in Table 1.

TABLE 1

Different lens parameters of the mouse strain HAN:NMR-I (♀) (data per lens)

CL = clear and transparent lens; MV = multiple vacuoles in the lens posterior cortex; % = the percentage of total proteins of the lens. Water-soluble Dry Weight = Water-soluble proteins = Crystallins; Water-insoluble Dry Weight = Water-insoluble proteins = Albuminoid.

Age (days)		30	90	112	257	566	793
Slit-lamp examination		CL	CL	CL	CL	MV	MV
Lens Fresh Weight (mg)	LFW	3.51	5.40	5.98	7.60	9.80	10.85
Lens Dry Weight (mg)	LDW	1.14	1.71	2.69[+]	3.92[+]	4.67	4.94
Water-soluble Dry Weight (mg)	WSDW	1.07	1.59	1.81	2.12	3.32	3.10
Water-insoluble Dry Weight (mg)	WIDW	0.07	0.12	0.15	0.37	1.35	1.84
Ratio WSDW/WIDW (mg/mg)	R	15.3	13.3	12.1	5.7	2.5	1.7
Number of lenses	n	8	8	8	4	2	2
Water-soluble Dry Weight (% of total protein)		93.9	93.0	92.3	85.1	71.1	62.8
Water-insoluble Dry Weight (% of total protein)		6.1	7.0	7.7	14.9	28.9	37.2

[+]These values are too high, perhaps due to variation in the biological material.

Methods

Thin-layer isoelectric focusing on polyacrylamide gel plates, protein fixation, washing out of the Ampholine carrier ampholytes, and staining were carried out as described by Bours (ref. 9). Because isoelectric focusing depends on uniform staining and destaining of the gels, it is generally postulated that precise and reliable staining with Coomassie Blue is proportional to protein weight (refs. 13-15).

Capillary isotachophoresis of the proteins was performed with an LKB 2127 Tachophor equipped with the standard capillary tube (23 cm x 0.45 mm I.D.), thermostatted at 12°C (ref. 1). Quantitative estimation of the protein fractions (monitored by UV absorption at 280 nm) was done by peak surface integration with a Spectra-Physics Minigrator (Model 602). This electronic integrator was programmed to hold its baseline from the start until the end of each run and to separate the protein peaks by verticals dropped at the lowest point between two peaks.

Chemicals

MES (2)N-morpholino)-ethanesulfonic acid, Merck N° 6128)was used as delivered. Ammediol (2-amino-2-methyl-1-3-propanediol, Merck N° 801464) was recrystallized from absolute ethanol and dried under vacuum at room temperature for 24 h. EACA (ε-aminocaproic acid, Merck N° 800145) was recrystallized from aqueous ethanol and dried as above. HPMC 15000 (hydroxypropylmethylcellulose, 15000 centipoise, Dow Chemical, Midland, Mich., U.S.A.) was dissolved at 1% (w/v) in glass-distilled water, in the cold room. After 24 h, the solution was transferred to dialysis tubing and dialyzed for 72 h against three changes of ten volumes of glass-distilled water. The dialyzed stock solution was kept in glass bottles at 4°C. Ampholine carrier ampholines (LKB), 2 pH unit ranges were used as delivered. The amino acids were of chromatographic quality.

Solutions

With minor modifications, the composition of the electrode solutions used for most experiments was based upon the work of Kjellin et al. (ref. 4) and of Everaerts et al. (ref. 16), and upon data furnished by LKB produkter. All solutions were made up in glass-distilled water that was freshly boiled for 15 min and cooled without access to air. The graduated flasks, containing weighed amounts of chemicals, were flushed with nitrogen just prior to adding the water. The finished solutions were immediately distributed into appropriate plastic syringes, with hermetically capped hubs and kept at 4°C until needed.

Leading electrolyte solution for rinsing and filling the capillary, 96 mg of MES, 106 mg of Ammediol, and 40 ml of stock HPMC solution. After making the volume up to 100 ml, this solution is rapidly filtered under vaccum through a 5mμ Millipore filter. The pH is 9.1

Leading electrolyte solution for the electrode vessel: 96 mg of MES and 111 mg of Ammediol made up to 100 ml (pH 9.1). (To obtain exactly the same pH in the two leading electrolytes, i.e. with and without HPMC, slightly different amounts of buffering counter-ions had to be used.

Terminating electrolyte solution: 148 mg of EACA and 106 mg of Ammediol are dissolved in 90 ml of water and brought to pH 10.8 with freshly prepared and filtered Ba(OH)$_2$ solution. Contact with atmospheric air must be kept to an absolute minimum during this operation. The solution is made up to 100 ml.

Spacer mobility gradient solution: Unless otherwise stated, the following solution was used: 0.4 ml of amino acid mixture (1 ml contained 4 mg of glycine, 4 mg of valine and 3.6 mg of β-alanine), 0.18 ml of Ampholine, pH range 7-9, 40 % (w/v), 0.095 ml of Ampholine, pH range 8-9.5, 40 % (w/v), and 0.3 ml of Ampholine, pH range 9-11, 20 % (w/v), made up to 10 ml with glass-distilled water.

General operating procedure

Before starting a series of experiments, the whole apparatus is thoroughly rinsed with distilled water. One of the 20-ml syringes containing leading electrolyte solution with HPMC is put into the left-hand receptacle, and the capillary is filled with this solution, taking care to push through several air bubbles interspaced with liquid. The leading electrolyte vessel is filled from above with leading electrolyte without HPMC. One of the 20-ml syringes, containing the terminating electrolyte, is connected and used as described in the instruction manual. The sample is injected about 2 mm inside the leading electrolyte. In order to achieve quantitative and reproducible sample injection, the appropriate volume of spacer solution (2 µl) is first withdrawn into the 10-µl syringe, and this is followed by the sample solution (3 µl of the 1 % lens protein solution). After injection, the microsyringe is slowly withdrawn, while giving it half turns in both directions.

Each run was started with the power supply in the constant voltage mode at 7 kV, which initially supplied   150 µA current. When the current dropped to 40 µA the power supply was manually switched to the constant current mode (usually 12-13 min after the start of the run). During detection, under a constant current of 40 µA, the voltage rose from about 8.5 kV to about 11 kV. Detection occured between 19 and 26 min after sample injection using a recorder chart speed of 5 cm/min.

Immediately after the end of an experiment, the capillary is rinsed with the leading electrolyte solution, preferably by pushing through several air bubbles interspaced with liquid. Before the start of each new separation experiment, the liquids in both electrode vessels are completely replaced. This eliminates possible irreproducibility due to either of two phenomena: (1) CO_2 absorption in the electrolytes, and (2) changes in pH and composition of the electrolyte solutions due to electrolysis.

At the end of the day the whole apparatus is first rinsed with water, and then the capillary is filled with a 1 % solution of Triton X-100. This solution can be left in the capillary overnight, but one must take care to rinse sufficiently before starting a run.

RESULTS

The normal mouse lenses were studied at six ages: 30, 90, 112, 257, 566, and 793 days. The life expectancy of this albino strain is 2.4 years (876 days = 80 % of the lethal full time; lethal half time = 722 days). The transparency was assessed by slit-lamp examination, and after enucleation wet and dry weights and water-soluble and water-insoluble dry weights were determined (Table 1). The wet weight values of the lenses from the albino mice from these experiments were expressed as the equation of a linear regression line: 5.15 log (age - 4.40 ($S_{y.x}$ = $\pm$ 0.38; r = 0.99). The values from the mouse lenses published by Zigman and co-workers (refs. 17 and 18), were expressed as the equation: 7.88 log (age) - 9.21 ($S_{y.x}$ = $\pm$ 0.73; r = 0.97). The ratio (R) of the water-soluble to the water-insoluble protein is shown in Fig. 1, and reveals an absolute decrease of the water-soluble and an absolute increase of the water-insoluble crystallins in the lens. The ratio R decreases markedly from 15 for a 30-day-old lens to 1.7 for a 793-day-old lens.

The thin-layer isoelectric focusing patterns of the 30- to 793-day-old mouse lens crystallins (Fig. 2) show distinct differences between the young and the old lenses. The γ-crystallins from the 30-day-old lens are prominent (Fig. 2b) and are separated into 8 individual components. The β-crystallins of high isoelectric point (pH 6.8-7.0) and low molecular weight are also quite prominent in this stage. In the 793-day-old lens (Fig. 2e) the concentrations of nearly all γ-crystallin components and of one β-crystallin component (N° 9) are

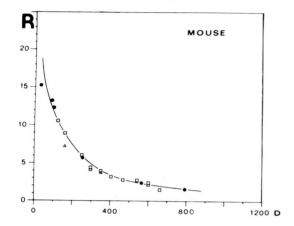

Fig. 1. The ratio R of the mouse lens is the quotient of the water-
soluble dry weight and the water-insoluble dry weight.
● = data from the present experiment; △ = data calculated
from Zigman et al. (ref. 17); □ = data calculated from Zigman
et al. (ref. 18). This curve approaches that of a hyperbola,
according to the formula y = $\frac{a}{x - b}$ + c where a = 875, b = 35
and c = 1.2.

reduced sharply. The following samples are used as references (1).
β_s-Crystallin preparation (Fig. 2a, f) having its main band at pI =
7.00 (ref. 19). (2) Bovine lens equator (25-year-old, Fig. 2g). Due
to aging of the lens equator this sample contains only very small
amounts of γ-crystallins at pI > 7.00 and serves as an additional
marker for the position of the last β-crystallin, β_s, at pI = 7.00
(ref. 19). (3) Bovine lens nucleus (25-year-old, Fig. 2h). This sample
is taken from innermost part of the lens which is the embryonic nucleus
and contains all γ-crystallins focused at positions > 7.00 (ref. 19).
The densitometric tracings (Fig. 3) clearly demonstrate the reduction
of γ-crystallin components N° 2, 3 and 8 and a loss of the components
N° 1, 4, 5, 6 and 7. This tracing also demonstrates the decrease in
the β-crystallin component 9 and a relative increase in the high-
molecular-weight crystallin components N°s 1-4 and 5-6.

The isotachophoresis experiments (Fig. 4) show somewhat higher
resolution of the β- and γ-crystallin components. The β-components
N°s 1-11 (β_I - β_{III}), which show a relative increase in aged lenses

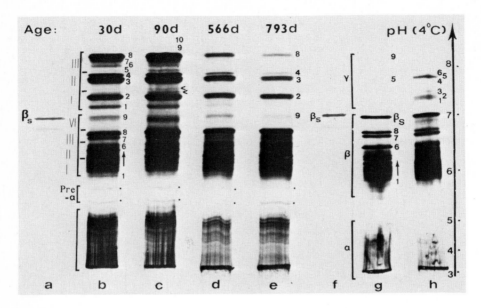

Fig. 2. Thin-layer isoelectric focusing patterns of the crystallins
of the mouse lens at various ages. The samples b-e and g-h
amount to 300 µg of protein a = f = fairly pure β_{slow}-crystallin
(ref. 20) of pI = 7.00; g = bovine lens equator (25-year-old);
h = bovine lens nucleus (25-year-old). The scale shows the pH
values along the gel, measured at 4°C. I- IV denote the groups
of β-crystallin components: I = β_{1-2}, II = β_{3-6}, III = β_{7-8},
IV = β_9. I - III denote the groups of γ-crystallin components:
I = γ_{1-2}, II = γ_{3-4}, III = $\gamma_{5-8(-10)}$. α = α-crystallins,
pre-α = pre-α-crystallin, β = β-crystallins, γ = γ-crystallins.

(Fig. 4d), are followed by the amino acid spacer glycine (which is a

broad zone without UV absorption) and the β-components NOS 12-14 (β_{IV}).

These low-molecular-weight β-crystallins preceded by glycine and

followed by valine decrease with age (Fig. 4) and are similar in

electrophoretic mobility to the low-molecular-weight β-crystallin (LMC)

isolated by Rabaey (ref. 21). The β-component N° 12 is similar in

electrophoretic mobility to the β_s-crystallin (ref. 20). Isotachophore-

tic analysis of this component is shown in Fig. 5(a-c). The groups

of crystallin species found by isotachophoresis are identical to the

crystallin species found by isoelectric focusing. The essential marker

in this respect is the β_s-crystallin, which appears at the same

position in the crystallin pattern with isoelectric focusing (Fig. 2a,

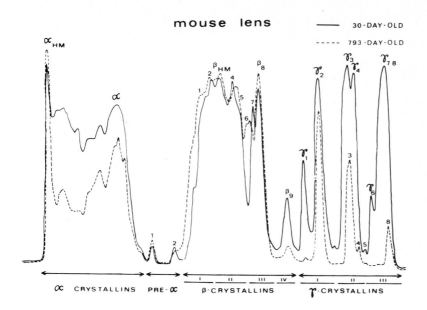

Fig. 3. Densitometric tracings of the crystallins of the 30-day-old
(————————) and 793-day-old (--------) mouse lens after isoelectric focusing. The tracings were made from the photograph
shown in Fig. 2. Each peak is marked with arabic numerals which
correspond to the components shown in Fig. 2. I - IV are
explained in the legend for Fig. 2.

f; g; h; 3) and with isotachophoresis (Fig. 5a,b). All γ-crystallin
components (N°s 1-16) decrease sharply. The decrease is particularly
marked for N°s 6, 9, 11, 14 and 15. The γ-crystallins components are
separated by the β-alanine ion into components N°s 1-12 (γ_I, γ_{II}, γ_{III})
and N°s 13-16 (γ_{III}). In the 793-day-old lens the concentration of pre-
α-crystallin also deceases slighthly during aging (Fig. 4a-d).

The amounts of the individual water-soluble crystallins have been
calculated as percentage water-soluble (% WS, Fig. 6a). Both isoelectric
focusing and isotachophoresis show a relative increase in the total
β-crystallin concentration which is accounted for by increase in the
β-crystallin groups I and II. The percentage concentration of γ-
crystallin in the soluble crystallins also decreases (Fig. 6a).

230

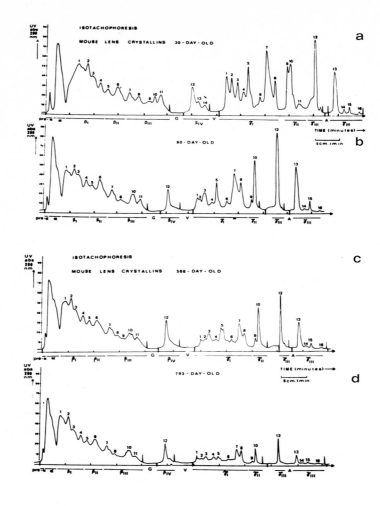

Fig. 4. The patterns of capillary isotachophoretic analysis of (a) 30-day-old mouse lens; (b) 90-day-old mouse lens; (c) 556-day-old mouse lens; (d) 793-day-old mouse lens. All samples amount to 30 µg of protein. I - IV denote the groups of β-crystallin components: I = β_{1-4}, II = β_{5-6}; III = β_{7-11} IV = β_{12-14}. I - III denote the groups of γ-crystallin components: I = γ_{1-8}; II = γ_{9-11}, III = γ_{12-16}. pre-α = pre-α-crystallin, α = α-crystallin; β = β-crystallin. G = glycine, V = valine, A = β-alanine.

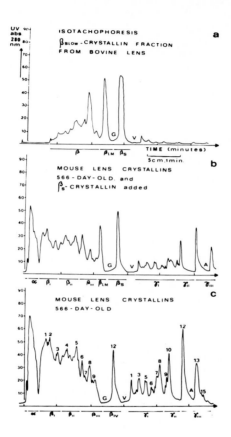

Fig. 5. Capillary isotachophoresis of a fairly pure β$_s$-crystallin fraction (ref. 20) from the bovine lens, compared with 566-day-old mouse lens. (a) β$_s$-crystallin fraction, 30 µg; (b) 10 µg of β$_s$-crystallin + 30 µg of 566-day-old mouse lens crystallins;(c) 30 µg of 566-day old mouse lens crystallins. For further explanations see Fig. 4.

It should be emphasized, however, that these percentages are only relative, due to the fact that the albuminoid increases continuously in the lens. When recalculated to account for albuminoid and soluble crystallins (Fig. 6b) a comparison of the results for the 30- and

232

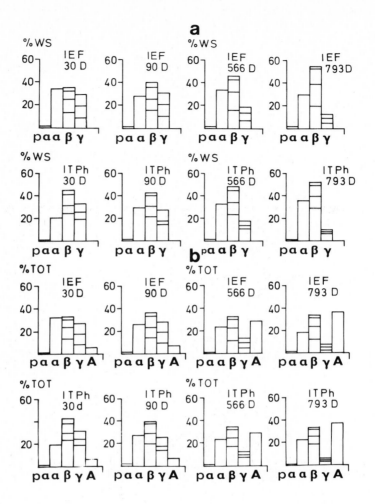

Fig. 6. The distribution of α-, β-, and γ-crystallins and their groups
of components in the mouse lens during aging, evaluated by
densitometry of thin-layer isoelectric focusing patterns (IEF)
and by a computing integrator system of isotachophoretic patterns (ITPh). The divisions in the columns representing the
β- and γ-fractions indicate the relative (a) or obsolute (b)
amounts of each subfraction that was obtained. These subfractions are, from bottom to top, I - IV for the β-fraction
and I - III for the γ-fraction, respectively. Further explanations are given in the legends of Figs. 2 and 4. (a) % WS =
percentage of water-soluble crystallins per lens; (b) % TOT =
percentage protein per lens (water soluble crystallins +
albuminoid). The computed peak surfaces obtained by isotachophoresis (Fig. 4a-d) are divided by the specific extinction
coefficients, taken from the rat lens: for α-crystallin $E^{1\%}_{1cm}$ =
8.0 (ref. 22), for β-crystallins $E^{1\%}_{1cm}$ + 18.6 (ref 23) anf for
γ- crystallins $E^{1\%}_{1cm}$ = 18.0 (ref. 22)

793-day-old lenses, as analyzed by two entirely different methods, isoelectric focusing and isotachophoresis, revealed the following, in absolute numbers:

    (1) The percentage of α-crystallins remains about the same (Fig. 6);

    (2) The percentage of the pre-α-crystallin shows a slight decrease with age (Figs. 3 and 4);

    (3) The percentage of the total amount of β-crystallins decreases. The β_I-crystallin decreases slightly, the β_{II}- and β_{III}- crystallins remain the same, and the β_{IV}-crystallin decreases considerably;

    (4) The number of γ-crystallin components per lens decreases.

DISCUSSION

Reproducible separation of lens crystallins by isoelectric focusing is possible, although the α-crystallins of high and low molecular weight precipitate due to low solubility at their isoelectric points. A second problem in analyzing crystallins by thin-layer isoelectric focusing is the difficulty in discriminating between the α-crystallins (both of low and of high molecular weight). With increasing age of the lens, the high molecular weight α-crystallin increases in concentration. On thin-layer isoelectric focusing, the high molecular weight α-crystallin hardly enters the polyacrylamide gel, remaining at the origin, while the low molecular weight α-crystallin precipitates and is visualized, after staining, as a streak.

With the capillary isotachophoresis technique described, the α-crystallins appear as a single peak and it is impossible to descriminate between the two α-crystallins. For the other crystallins a very good separation into a large number of components (at least 32) has been obtained by means of Ampholine carrier ampholytes with amino acids as spacer ions. The time required for one analysis is 20 to 30 min. Short analysis time, small sample required, and high resolution and reproducibility make the capillary isotachophoresis a very useful technique for analysis of lens crystallins. Capillary isotachophoresis of serum proteins and CSF proteins has been shown to be of interest for the diagnosis of neurological diseases (ref. 1). Preliminary results of isotachophoresis studies of mouse lens crystallins were given and the overall reproducibility of the separation of these complex lens protein mixtures by this technique was demonstrated (ref. 1). Preliminary notes on the aging of mouse lens crystallins have been published by Bours (refs. 24 and 25).

The experimental conditions used in this study were selected to maximize the separation of the γ-crystallins, which were our primary interest. By choosing the appropriate pH range the resolution of the α-crystallins could similarly be enhanced, however, this would have simultaneously decreased the resolution of the γ-crystallins.

Pre-α-crystallin, which appears in isoelectric focusing between the α- and β-crystallin groups (ref. 26), exhibits a molecular weight of 42.000 D (ref. 27). It is one of the β_L-crystallins, as shown by gel filtration (ref. 28) and by electrophoresis (ref. 27).

In isotachophoresis this pre-α-crystallin appears to precede α-crystallin (Fig. 4). This is due to the fact that pre-α-crystallin changes from an isoelectric to a charged molecule in the isotachophoresis system. Similar changes in positions have been established for other molecules, including serum proteins and hemoglobin (cf. Arquembourg, ref. 29). The molecular weight of pre-α-crystallin has been erroneously determined as 14,000 D by Van den Broek et al. (ref. 30), possibly indicating the existence of subunits in SDS (ref. 27).

Whereas the low-molecular-weight β-crystallins (β_{IV}) and the γ-crystallins decrease markedly with aging, the β-crystallins of high molecular weight and low isoelectric point (β_I) decrease slightly or remain about the same (Fig. 6b). Simultaneously, the amount of albuminoid increases absolutely with aging. The latter has been demonstrated in the mouse lens by Dapson et al. (ref. 31) and by Zigman et al. (refs..17 and 18). However, the extent to which the increase in albuminoid (Table 1) is related to the decrease of β_{IV}- and γ-crystallins, is not known.

Several authors have studied the aging of the mouse lens by analysis of lens crystallins at different stages on the growth curve. Because Day and Clayton (ref. 32) found no major differences in crystallin subunit composition between six mouse strains, we felt it allowable to compare our results with those from several other authors. The total α-, β-, and γ-crystallins that we determined by isoelectric focusing were compared with the results obtained by others using polyacrylamide gel electrophoresis. These data were recalculated either as percentage of water-soluble protein or as percentage of total protein (including albuminoid) and are given in Table 2. Table 2A demonstrates that of

TABLE 2

Concentration of crystallins and albuminoid in whole mouse lenses of different strains (collected data from electrophoresis and isoelectric focusing experiments)

-- = data incomplete; S_{yx} = the standard deviation of the linear regression line; r = the appropriate correlation coefficient.

Age (days)	Concentration of water-soluble protein (1)			Concentration of total protein (2)				ref.
	Crystallins			Crystallins			Albuminoid	
	α-	β-	γ-	α-	β-	γ-		
30	34.4	36.3	29.3	32.3	34.1	27.5	6.1	This article, Figs.2,3,6
49	35.5	21.1	43.4	34.1	20.2	41.5	4.2	(17)
90	28.1	41.1	30.8	26.2	38.2	28.6	7.0	This article, Figs.2,3,6
119	42.2	33.4	24.4	38.6	30.5	22.2	8.7	(18)
161	37.2	36.1	26.7	32.7	31.7	23.4	12.2	(18)
245	32.8	44.1	23.1	28.2	37.9	19.9	14.0	(18)
294	33.6	38.3	28.1	27.1	30.8	22.6	19.5	(18)
294	30.2	49.1	20.7	24.6	39.9	16.9	18.6	(18)
350	32.0	48.7	19.2	25.5	38.7	15.4	20.4	(18)
350	35.1	39.3	25.6	27.8	31.2	20.4	20.6	(17)
406	27.7	--	--	21.1	--	--	23.7	(17)
469	36.2	--	--	26.8	--	--	26.0	(17)
566	33.5	47.7	18.8	23.8	33.9	13.4	28.9	This article, Figs.2,3,6
602	21.0	--	--	14.5	--	--	30.9	(18)
602	25.7	44.1	30.2	18.3	31.3	21.5	28.9	(18)
658	25.8	45.4	28.8	15.6	27.4	17.5	39.5	(18)
793	30.0	57.0	13.0	18.8	35.8	8.2	37.2	This article, Figs 2,3,6
S_{yx}	5.31	5.97	6.29	3.58	5.22	5.32	2.02	This article, Figs. 2,3,6 (17,18)
r	-0.57	+0.75	-0.54	-0.86	+0.09	-0.76	+0.98	
S_{yx}	3.61	2.69	1.47	2.83	2.38	1.48	1.08	This article, Figs. 2,3,6
r	-0.08	+0.97	-0.99	-0.91	-0.21	-0.99	+0.99	

(1) The concentration is expressed as a percentage of the soluble protein

(2) The concentration is expressed as a percentage of the total protein (water-soluble crystallin and albuminoid). These data are recalculated from the original tables taken from various authors by incorporation of the percentage of albuminoid.

the water-soluble components, α- and, more markedly, γ-crystallins decrease, whereas β-crystallins increase. By contrast, when one takes the albuminoid into consideration (Table 2B), all the crystallins decrease. This decrease is most pronounced for γ-crystallin. On the other hand, the albuminoid increases rapidly in such a way that, at 793 days of age, about 40 % of the mouse lens proteins is albuminoid (Table 2B). The amount of albuminoid increases steadily and linearly with age. The linear regression line for albuminoid has small standard deviation and a high correlation coefficient (r = 0.98; Table 2B). This result shows that our data and that of Zigman et al. (refs. 17 and 18) correspond very well (Table 2). Also the values for the γ-crystallins in table 2 correlate fairly well (r = 0.76). The correlation coefficients in Table 2 for α- and β-crystallins are low, due to the fact that these concentrations follow the growth curve of the lens and cannot be filled to a straight line such as we have just described for the decrease of γ-crystallin and the increase of albuminoid.

CONCLUSIONS

(1) For the analysis of crystallins a direct comparison between thin-layer isoelectric focusing and capillary isotachophoresis is possible and useful. Nearly all the components have the same sequence from anode to cathode in both methods.

(2) Because separation is better and the results are also quantitative, the information gained from analytical isotachophoresis is greater than that obtained from isoelectric focusing.

ACKNOWLEDGEMENTS

This research was supported by the Deutsche Forschungsgemeinschaft (Mo. 249/8; Oh 32/1.).

REFERENCES

1 P. Delmotte, Sci. Tools , 24(1977)33-41.
2 L. Arlinger , Biochim. Biophys. Acta 393(1975)369-403.
3 L. Arlinger, in Righetti (Editor), Progress in Isoelectric Focusing and Isotachophoresis. North-Holland, Amsterdam, 1975, pp. 331-340.
4 K.G. Kjellin, U. Moberg and L. Hallander, Sci. Tools, 22(1975)3-7.
5 K.G. Kjellin, L. Hallander and U. Moberg, J. Neurol. Sci., 26(1975) 617-622.
6 A. Kopwillem, W.G. Merriman, R.M. Cuddeback, A.J.K. Smolka and M. M. Bier, J. Chromatogr., 118(1976)35-46.
7 U. Moberg, S.G. Hjalmarsson, L. Arlinger and H. Lundin, in. Radola and Graesslin (Editors) Electrofocusing and Isotachophoresis, De Gruyter, Berlin, New York, 1977, pp. 515-526.

8 P. Delmotte, in Radola and Graesslin (Editors) Electrofocusing and Isotachophoresis, De Gruyter, Berlin, New York, 1977, pp. 559-564.
9 J. Bours, in Radola and Graesslin Electrofocusing and Isotachophoresis, De Gruyter, Berlin, New York, 1977, pp.303-312.
10 F.M. Everaerts, J.L. Beckers and Th. P.E.M. Verheggen, Ann. N.Y. Acad. Sci., 209(1973)419.
11 J. Bours, Interdiscipl. Topics Geront., 12(1978)196-204.
12 O. Hockwin, U. Bechtel-Ehrig, W. Licht, E. Noll and F. Rast, Ophthal. Res., 2(1971)77-85.
13 W.N. Fishbein, Anal. Biochem., 46(1972)388-401.
14 J. Bours and J. van Doorenmaalen, Sci. Tools, 17(1970)36-38.
15 J. Bours, Sci. Tools, 20(1973)29-34.
16 F.M. Everearts, J.L. Beckers and T.P.E.M. Verheggen, Isotachophoresis - Theory, Instrumentation and Applications, Elsevier, Amsterdam, 1976.
17 S. Zigman, J. Schultz and T. Yulo, Exp. Eye Res., 15(1973)201-208.
18 S. Zigman, T. Yulo and J. Schultz, Ophthal. Res., 6(1974)259-270.
19 J. Bours, 16(1973)501-515.
20 A.F. van Dam, Exp. Eye Res., 5(1966)255-266.
21 J. Bours and M. Rabaey, Exp. Eye Res., 20(1975)180-181.
22 S. Lerman, Can. J. Biochem., 47(1969)1115-1119.
23 C.R. Hamlin, Exp. Geront., 4(1969)189-195.
24 J. Bours, Exp Eye Res., 25(1977)530.
25 J. Bours, Ophthalmologica, (Basel), 176(1978)281-282.
26 J. Bours, in Righetti (Editor) Progress in Isoelectric Focusing and Isotachophoresis, North-Holland, Amsterdam, 1975, pp. 235-256.
27 G.A. Griess, S. Zigman and T. Yulo, Mol. Cell. Biochem., 12(1976) 9-14.
28 J. Bours and S.K. Brahma, Exp. Eye Res., 16(1973)131-142.
29 P.C. Arquembourg, Immunoelectrophoresis: Theory, Methods, Identifications, Interpretations, Karger, Basel, 2nd ed., 1975.
30 W.G.M. van den Broek, J.N. Leget and H. Bloemendal, Biochim. Biophys. Acta 310(1973)278-282.
31 R.W. Dapson, J.G. Otero and W.R. Holloway, Nature (London, 218 (1968)573.
32 T.H. Day and R.M. Clayton, Biochem. Genet., 8(1973)187-203.

A. Adam and C. Schots (Editors), *Biochemical and Biological Applications of Isotachophoresis*   239
© 1980 Elsevier Scientific Publishing Company, Amsterdam — Printed in The Netherlands

ISOTACHOPHORESIS OF CEREBROSPINAL FLUID PROTEINS

K.G. Kjellin and L. Hallander
Department of Neurology, Karolinska Hospital, Stockholm (Sweden)

SUMMARY

Isotachophoresis (ITP) was applied to cerebrospinal fluid (CSF) by Kjellin et al. (Sci. Tools, 22,(1975)3; J. Neurol. Sci., 26(1975)617) who described the advantages of this new technique for studying CSF- and serum proteins in patients with neurological diseases, e.g. multiple sclerosis (M.S.). ITP of CSF and serum proteins has been performed both in capillary tubes and in simple plastic tubes with polyacrylamide gels, and different analytical systems have been reported since 1975. The ITP findings have been compared with those of electrophoresis and isoelectric focusing. A microdialysis procedure has been elaborated for CSF prior to ITP. A cationic ITP system is preferred for examining samples from patients with abnormalities in the gamma-globulin region, e.g. MS and infectious diseases. The ITP fractions are better separated in a two-turn capillary tube plate. The proteins can be identified by using "internal standards" or better by a micropreparative immunological procedure.

ITP systems used are described as well as the results from control cases and patients with neurological diseases, mostly MS.

PLENARY SESSION LECTURE

In 1975 Kjellin et al. described the application of isotachophoresis (ITP) to cerebrospinal fluid (CSF). The authors reported the advantages of this new technique for studying proteins of the CSF and serum in neurological diseases such as multiple sclerosis (M.S). Very small spinal fluid samples, a few μl of concentrated and 15-30 μl of uncon- centrated samples, could be analysed by the capillary tube technique used. The examinations were performed rather quickly - in about 30-60 min - with the results immediately obtained on a recorder. CSF generally with a very low protein content, is "self-concentrated" by ITP so that unconcentrated CSF can be examined, avoiding losses by prior concentration procedures such as ultrafiltration. Furthermore, low-molecular-weight compounds such as those in CSF ultrafiltrates can be analysed as well as high-molecular-weight compounds including proteins. The ITP findings were compared with those of electrophoresis and isoelectric focusing. The technique was found to give high resolution, reproductible results, and was easy to perform.

Since 1975 the authors have reported investigations on proteins in CSF and serum where ITP has been performed both in capillary tubes and in simple plastic tubes with polyacrylamide gels (see the following paper for references). Various ITP systems have been described. Anionic as well as cationic systems have been elaborated. Mostly anionic ITP systems have been used for separating acid proteins, and cationic systems for alkaline proteins. The general intention is to perform the ITP as close to physiological conditions as possible, in order to preserve the proteins. Thus very acid and alkaline pH regions are avoided as well as too low terminating conductivities. ITP in more restricted mobility ranges, using a two-turn LKB Tachophor capillary tube, was performed to achieve higher resolution and larger loading capacity.

A microdialysis procedure (Fig. 1) has been developed to precede the ITP. The dialysis removes low-molecular-weight substances such as NaCl (CSF is indeed rather similar to physiological saline); this is important to obtain faster as well as more comparable results since some low-molecular-weight substances in biological samples obviously behave as internal spacers. Furthermore, the limiting proportion between amount of CSF and ITP capillary length, when the protein in larger CSF volumes are incompletely separated due to salt overloading, is overcome (see the following paper). Contrary to microdialysis through collodion membranes where losses of alkaline proteins, espe-cially of immunoglobulins, were found, the polyacrylamide gel dialysis seems to give almost complete recovery also of low-molecular-weight proteins. The microdialysis procedure used is described in detail and discussed in the next paper.

CSF and blood samples are collected in a standardized way: fasting, midday. After centrifugation to remove corpuscular elements, samples of CSF and corresponding serum are analysed. The ITP results are

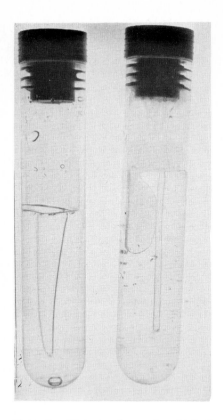

Fig. 1. The equipment used for microdialysis is polyacrylamide gel.
        A detailed description is given in the following paper
        (Hallander and Kjellin).

compared with those of electrophoresis and isoelectric focusing as
mentioned previously. The centrifuged samples are stored for not
more than a few days at +4°C and are otherwize frozen at -23°C.

ITP has been performed on samples from control cases, patients
with plasma cell dyscrasia, and neurological diseases, especially
disorders with aberrations in the gamma-globulin region of the CSF
such as MS. An example is shown in Fig. 2, where a cationic system
in capillary ITP has been used for separating CSF proteins from a
control case and a patient with MS.

In a limited number of cases immunological identification tests
have been performed by a micropreparative procedure using a LKB
Tachophrac apparatus. Immunological identification is of course more
significant than using internal marker proteins.

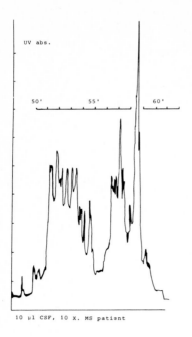

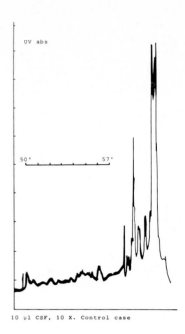

10 µl CSF, 10 X. MS patient   10 µl CSF, 10 X. Control case

Fig. 2. The UV diagrams of CSF proteins from a control case and a
patient with MS, separated by a cationic system in capillary
ITP. Leading electrolyte: Tris 5 mM, cacodylic acid 20 mM,
pH 5.5. Terminating ion: ß-alanine. Spacers 3 µℓ of 2 %
ampholine pI "6.5-8". Creatinine 1 µℓ, 25 mM. Samples: According to figure; not dialysed. Note the difference in immunoglobulin profiles between the two diagrams, due to the high
concentration of IgG, especially with more alkaline pI, in
the CSF from the MS patient.

In conclusion, ITP as reported by the authors in several communications since 1975, has been found to be a very versatile method with
high separation of CSF proteins. The anionic and cationic systems
developed as well as a microdialysis procedure in polyacrylamide gel,
are described in detail and discussed in the following paper including
the references. The clinical advantages of ITP are obvious; these
include the need for only very small samples, and fast and reproducible
results by a technique easy to perform for a given analytical system.
ITP can be combined with other methods for investigating the CSF
proteins; for example, the various staining techniques and immunological identification tests such as the micropreparative procedure
used by the authors for fractions separated by capillary ITP.

ACKNOWLEDGEMENTS

The investigations summarized in the plenary session lecture presented above and those reported in the following paper, were supported by grants from the Karolinska Institute and the Swedish Society of Medical Sciences.

REFERENCES

See following paper.

A. Adam and C. Schots (Editors), *Biochemical and Biological Applications of Isotachophoresis*   245
© 1980 Elsevier Scientific Publishing Company, Amsterdam — Printed in The Netherlands

ISOTACHOPHORETIC SYSTEMS FOR STUDYING PROTEINS IN CEREBROSPINAL FLUID
AND SERUM

L. Hallander and K.G. Kjellin
Department of Neurology, Karolinska Hospital, Stockholm (Sweden)

SUMMARY

Various isotachophoretic (ITP) systems for investigating proteins
in cerebrospinal fluid (CSF) and serum have been reported by the
authors since 1975. The analytical systems most used are presented.
Cationic systems are described where in particular the effect of
creatinine has been evaluated. An anionic ITP system has been elaborated: leading electrolyte acetic acid 10 mM, Bistris 10 mM, pH 5.5;
terminating ion 4-(2-hydroxyethyl)-1-piperazineethanesulfonic acid
(HEPES). The separations are performed in a two-turn capillary tube
plate (LKB Tachophor) with 2 µl of 2 % Ampholine ^{R}(pI4.5 -5.0) added.
The system stacks from albumin to transferrin.

An apparatus for microdialysis of CSF in polyacrylamide gel is
described, including tests performed on radioiodinated proteins.

INTRODUCTION

Isotachophoresis (ITP) for the separation of proteins in cerebrospinal fluid (CSF) and serum with various anionic and cationic systems
has been performed in our laboratory during the last five years, as
described in our preceding paper in the plenary session.

Among the anionic systems used, the MES-Ammediol pH 9 system
(ref. 1) has been abandoned by the authors, since deposition of
proteins occurs at the capillary wall, which often gives "smoothed out"
UV-diagrams as well as "tripping".
These effects could be avoided if the capillary was carefully cleaned
every third separation. Another anionic system is reported by Kjellin
and Hallander (ref. 2).

The purpose of the present paper is to report some anionic and cationic ITP systems found suitable for CSF and serum-protein separations. In addition, a microdialysis procedure is described which has been developed for desalting small volumes (a few µl to some ml) of biological fluids prior to ITP. This procedure will probably replace the CSF ultrafiltration previously used by the authors. Contrary to a CSF dialysis technique described earlier (ref. 4), the present method is based on polyacrylamide gel diffusion. This technique was chosen because of its low protein adsorption and variable exclusion limits for molecules of different sizes.

MATERIALS

The methods were worked out for protein examinations of CSF and serum. The samples were collected as described in our foregoing paper. Capillary isotachophoresis (C-ITP) was performed in a LKB 2127 Tachophor, the capillaries being thermostatted at 10°C. Polyacrylamide ITP (PAG-ITP) in tubes was performed in a laboratory built disc electrophoresis apparatus. PAG-ITP in slabs was performed in a gel electrophoresis apparatus, GE 4, from Pharmacia Fine Chemicals, Uppsala, Sweden.

Acetic acid, ammonium persulphate, gamma-aminobutyric acid, glutamic acid, glycine, imidazole and tris were of pro analysi grade (Merck, Darmstadt, G.F.R.). The radioiodinated proteins, hPl (human placental lactogen) and IgG, as well as cytochrome c were a generous gift from Assistant Professor, P. Eneroth, Karolinska Hospital Stockholm, Sweden. Other chemicals were obtained as given in Table 1.

TABLE 1

List of Chemical and biochemical products.

Acrylamide	pro analysi, Merck, Darmstadt, G.F.R.
β-Alanine	
Histidine	
Taurin	
Ampholine^R	LKB-produkter, Bromma, Sweden.
Ammediol	Sigma. New York, U.S.A.
Bistris	
MES	
TES	
Cacodylic acid	Pfalz and Bauer, New York, U.S.A.
Creatinine	Hopkin and Williams Great Britain.
TEMED	Eastman Kodak, U.S.A.
Epsilon-aminocaproic acid (EACA)	Roth, Karlsruhe, G.F.R.
HEPES	BDH, Poole, England.
Bis-N-N-Methyleneacrylamide	(recrystallised before used).

METHODS AND RESULTS

Anionic systems:

The anionic MES-Ammediol system previously used has been replaced by other systems for reasons mentioned above. Acetate of glutamate (Glu) (pH 4.65-5.5) was chosen as leading ion of the new systems. The counter-ions, made up of one substance or a mixture or two, included tris, imidazole and bistris. The counter-ions preferred were those that gave less alkaline separation pH values.

In anionic PAG-ITP, tris was tried as counter-ion,(see ref. 2). The system TES 20 mM, tris 20 mM with ε-aminocaproic acid as terminating ion was developed for γ-globulins. Attempts to resolve proteins faster than hemoglobin with different spacer mixtures, leading ions and pH values were not sufficiently satisfactory with tris as counter-ion. However, with imidazol as counter-ion some progress was made (Fig. 1). Methods using vertical gel slabs are now being developed in which hemoglobin, for instance, can be used as "near terminator" marker protein (Fig. 7), migrating just ahead of taurin. The proper choice of terminating ion in ITP is more important in gel slabs than in gel tubes, because a straight "termination line" as well as a "sharp front" is needed for computerized comparative densitometric evaluation of the separation. In the system used hithertho beta-alanine was tolerable as terminating ion in gel tubes (Fig. 1) but glycine or taurin had to be used in gel slabs (Fig. 2). In tubes as well as in slabs, 5 μl of serum or 200-500 μl of CSF are needed to detect proteins after Coomassie Brilliant Blue R staining. Both modifications need 2-4 hours to be completed without prior dialysis of CSF.

In anionic C-ITP some systems were tried on a two-turn Tachophor[R] capillary, mainly using serum as a test mixture. The leading ion was acetate (10 mM acetic acid) with bistris as counter-ion. At pH 5.1 many peaks were obtained (Fig. 3) not belonging to ampholines. Various "pure" proteins showed "microheterogeneity" as multiple peaks. At pH 5.5 fewer peaks were obtained, but most proteins seemed homogeneous (Figs. 4-6).

Separation time was surprisingly short, taking into account the long migration distance compared to a straight capillary. Total separation time varied between 35 and 50 minutes, depending on the system and terminating ions (TES or HEPES). Most charge transport occurred

248

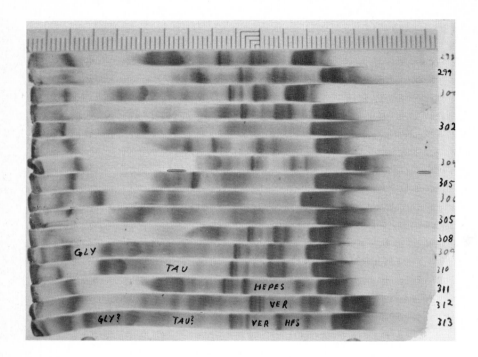

Fig. 1. Leading electrolyte: Glu 50 mM, imidazole 50 mM, (pH 5.5).
Terminating ion: beta-alanine. Sample: 5 µl of serum.
In gel 299 and 308, ampholines 40 % : 10 µl pI 5-7, 15 µl
pI 6-8. (= standard mixture). In gels 309-312: standard
mixture + 200 mM discrete spacers according to figure: glycine
15 µl, taurin 30 µl, HEPES (HPS) 30 µl veronal 30 µl. In gel
313: standard mixture + all discrete spacers combined. In
gels 305-308: overloading by ampholine mixtures. Run: 4 mA,
5 minutes; 8 mA, 45 minutes; 16 mA, 250 minutes. Metering
scale in mm.

within 20-30 minutes at 150-70 µA. Detection was at 50 µA for 10 min.
Driving the sample in and out of the injection part of the Tachophor
amounted to 2-10 min at 50 µA. Obviously, separation parameters change
rapidly close to pH 5, which is not far from the pK_a of acetic acid.

Cationic systems

Various cationic ITP electrolyte systems have been tried during
the last few years. They have all been based on tris as leading ion
(at pH 5.5-7.0), but the counter-ion and the terminating ion have

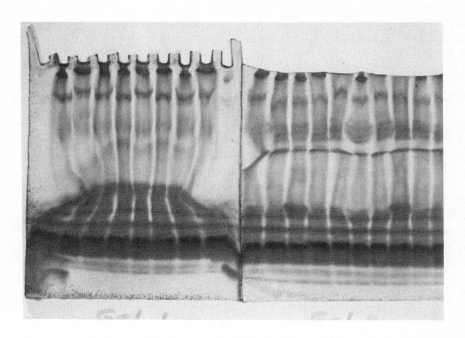

Fig. 2. Leading electrolyte: Glu 10 mM, counter-ions (tris:imidazole =
        1:7) 7.5 mM (pH 4.65). Spacers totally, ampholines 40 %: 50 µl
        pI 5-7, 50 µl pI 6-8. In gel 2 400 µl of 200 mM taurin were
        added.
        Sample totally: 200 µl of serum 50 % v/v in glycerol, traces
        of hemoglobin in some slits. Running time: 160 minutes, V_{max}
        400 V, W_{max} 10 W. I_{max} 40 mA. Original gel dimensions according
        to long gradient gels from Pharmacia Fine Chemicals.

been widely changed. Cacodylic acid was preferred as counter-ion in
C-ITP because of its pK_a value (6.21), good solubility and resistance
to microbes. Cationic PAG-ITP, where Glu was chosen as counter-ion
because of its pK_a, will be described elsewhere.

    The different cationic C-ITP systems elaborated have recently been
reported in detail by Kjellin and Hallander (ref. 3). As spacers,
different mixtures of Ampholine[R] in intervals of 0.5 pI were tested.
The pI "6.5-8" mixture was preferred, as it gave the best resolution
of IgG (see Fig. 2 in the foregoing paper) without "smoothing out"

250

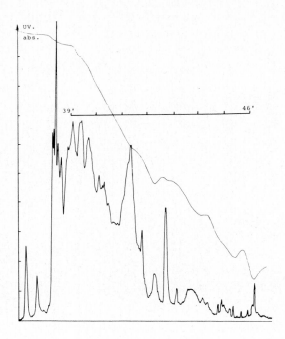

Fig. 3. Leading electrolyte: acetic acid 10 mM, bistris 7.5 mM (pH 5.1). Terminating ion: TES. Spacers: 2 µl of 1 % ampholine pI 5-5.5, 2 µl of 1 % ampholine pI 5.5-6. Sample: 1 µl of serum. Two-turn Tachophor capillary.

the UV curve by dilution. The acid proteins, most of them found to include transferrins by addition of standard solutions, were also sufficiently separated.

Different discrete spacers for cationic C-ITP were examined. The spacers migrated in the order bistris, histidine, Crea, EACA, γ-amino-butyric acid, β-alanine and glycine. All immunoglobulins (IgG) migrated ahead of Crea, but after bistris. Cytochrome c was used as a front marker. The rest of the proteins detected, except albumin, migrated between Crea and β-alanine. Albumin migrated between β-alanine and glycine if the separation took place under mild conditions (I_{max} =

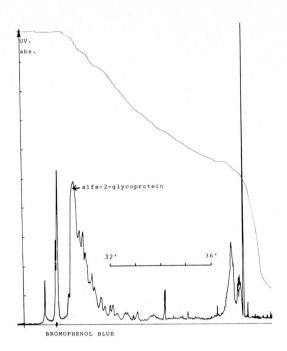

Fig. 4. Leading electrolyte: acetic acid 10 mM, bistris 10 mM (pH 5.5).
Terminating ion: HEPES. Spacers: 2 µl of 2 % ampholine pI
4.5-5. Sample: 10 µl of 0.1 % acid α-2-glycoprotein (dialyzed).
Two-turn Tachophor capillary.

50 µA), otherwise albumin coagulated, resulting in a "trip". The
albumin "peak" was asymetric (tailing) and distorted by numerous
spikes, probably coagulated solid matter.

When Crea was injected together with serum or CSF, or injected
5-10 min later, the proteins following IgG, i.e. all those following
Crea in the steady-state stack, appeared "magnified". This effect may
be a result of some kind of adsorption (Fig. 7). When a mixture of
IgG, Crea, transferrin and ampholines was injected, the transferrins
were not "magnified".

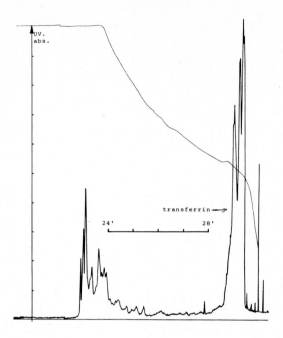

Fig. 5. Conditions as in Fig. 3. Sample: 1 µl of 2 % human serum
transferrin.

Microdialysis

Microdialysis of samples was found to be a valuable prior to ITP
as well as ioselective focusing. In ITP faster separation and larger
capacity (repeated sample application) is made possible. In isoelectric
focusing, band curvature is minimized.

C-ITP in a short, straight capillary with injection of more than
15-20 µl of CSF will give incomplete resolution even after long
separation times (45 min - 2 h), due to salt overloading. The
"smoothing out" of the UV diagrams is proportional to the amount of
salt injected. The same phenomenon was observed when serum diluted
in physiological saline (1:50, 1;100, 1:200) was injected in the same
amounts. Proteins, diluted in weak leading or terminating buffers,
were separated in normal time (25-50 min).

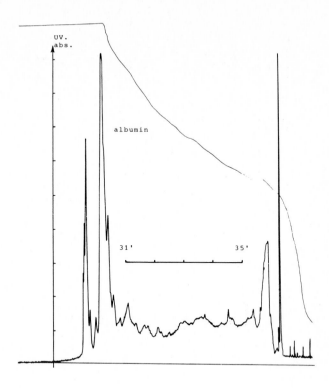

Fig. 6. Conditions as in Fig. 3. Sample: 1 µl of 2 % human serum
    albumin.

Microdialysis through collodium membranes resulted in losses of
alkaline proteins, especially immunoglobulins. Therefore polyacrylamide
gel dialysis was tried. Various polyacrylamide gel diffusion tests were
performed on CSF and serum, diluted in physiological saline as described above. The following parameters were found optimal to dialyse up to
a maximum 25 µl sample: a 3 ml $T_{20}C_8$ gel for 30-45 min at room temperature. No protein losses could be observed under these conditions, and
various stain solutions diffused readily. Leading and terminating
buffers diffused overnight into partially dried gels at room temperature.

Dialysator moulding in plastic 3 ml tubes (see Fig. 1 in the foregoing paper) was performed with 5.10^{-4} v/v TEMED (5 µl in 10 ml final
gel solution), and ammonium persulphate in sufficient amount to have a
gel meniscus after 15-20 min (max. conc. 10^{-3} w/w). Conical (150 µl)

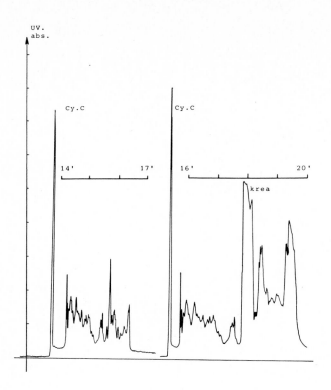

Fig. 7. Leading electrolyte: Tris 5 mM, cacodylic acid 7.5 mM (pH 6.3).
Terminating ion: β-alanine. Spacers: 1 μl of 1 % ampholine
pI 6.5-7, 1 μl of 1 % ampholine pI 7-7.5, 0.5 μl of 10 mM
bistris. Sample: 1 μl of serum, (creatinine added: 0.5 μl of
10 mM).

and straight (25 μl) dialysators were made. The inner core can be made
of most non-wettable plastic materials, not glass, silicone rubber or
metal. The dialysators were polymerized for 1 h at room temperature,
and heated in distilled water at 45-60°C for 2 h, in order to achieve
complete polymerization and persulphate decomposition and diffusion
out of the gel. When not used immediately, the gels could be stored
in 40-50%v/v methanol. The optimal dialysis time in conical dialysators
was 60 min, and in straight ones 30 min. Also 5 ml cones with 2 mm
thick walls have been used (dialysis time was 2-10 h).

The low diffusion of two radioiodinated proteins, human placental
lactogen (hPl) (M = 20 000) and IgG (M = 150 000), into the $T_{20}C_8$-gel

matrix is shown in figures 8 and 9. All dialysators were filled with
15 µl of protein solution (2000 cpm). After diffusion, dialysators

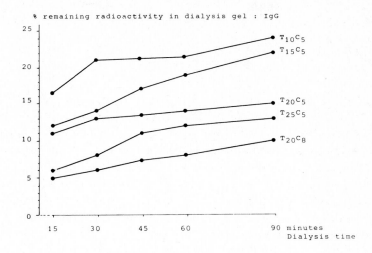

Fig. 8. Diffusion of radioiodinated IgG into polyacrylamide gel with
no salts added. (T = total acrylamide concentration % w/w.
C = fraction of crosslinker in gel % w/w ("bis")).

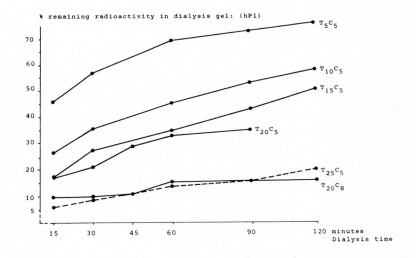

Fig. 9. Diffusion of radioiodinated human placental lactogen, into
polyacrylamide gel, with no salts added.

were washed with 5 x 15 µl of distilled water and the washings were
found to contain practically no radioactivity. The recovery after half
hour of dialysis for the $T_{20}C_8$ gel was found to be 90-95 %.

COMMENTS

The ITP methods presented in this paper have been designed to separate primarily proteins in CSF and serum, and will now be used for
investigation of neurological disorders.

As a general rule acid proteins will be separated in anionic systems,
and alkaline proteins in cationic systems. The pH of ITP will be around
neutrality, the proteins not migrating far from their natural pH range.
This means that important properties depending on pH, e.g. quarternary
structure, subunit aggregation and enzyme or receptor activity , are
less prone to irreversible damaged. Conceivably such methods as
zymogram techniques after ITP will be easier to perform, as well as
preparative ITP of labile proteins.

The need for less heterogeneous spacer mixtures was evident when
performing C-ITP in the two-turn capillary. Ordinary ampholines, even
in 0.5 pI ranges, readily caused dilution of the samples by creating
mixed zones. Ampholines with ranges of 0.1 pI or less gave better
results. However, the authors would prefer oligometric series of defined
pI and pK_a, something like biological buffers (ref. 5).

Desalting prior to ITP means that charge transport of no interest
is avoided, and instrumentation and time can be adequately exploited.
If dialysis is chosen as the desalting method, solvating agents can
be added to the sample simultaneously, and the correct starting pH for
the ITP achieved.

Initially, microdialysis of some hundred µl or less was performed
on collodium membranes. However, ITP and isoelectric focusing revealed
losses due to adsorption of alkaline proteins with this dialysis technique. This disadvantage was not observed in polyacrylamide gel dialysis, which is of importance in studying CSF proteins, where the ratio
of protein amount to dialysis area is small, and adsorption can disturb
the investigation to a greater extent than for serum proteins.
Furthermore, lower crosslinking and total acrylamide content, combined
with different diffusion time, will probably give information about
molecular weight.

The possibility of interaction between sample and components of the ITP system will be pointed out e.g. by π-orbital (aromatic), lipophilic or charge-complexing interactions.

Phenomena like Crea effect (Fig. 7) may in the future prove to be useful in studying interaction between macromolecules, such as enzymes, and small molecules, such as substrates or inhibitors. An "affinity lable ITP" may be developed, based on synthetic, easily detected, co-migrating or immobile "probes".

ACKNOWLEDGEMENTS

The investigations reported in this paper were supported by grants from the Karolinska Institute and the Swedish Society of Medical Sciences.

The skilful technical assistance of Miss Gunborg Strand and Mrs. Vera Snikvalds is gratefully acknowledged.

The authors are greatly indebted to Assistant Professor Peter Eneroth who generously allowed us the use of his LKB Tachophor apparatus.

REFERENCES

1 K.G. Kjellin, L. Hallander and U. Moberg, J. Neurol. Sci., 26(1975) 617-622.
2 K.G. Kjellin and L. Hallander, J. Neurol., (1979) in press; and references cited therein..
3 K.G. Kjellin and L. Hallander, J. Neurol., (1979) in press; and references cited therein.
4 P. Delmotte, In B.J. Radola and D. Graesslin '(Editors), Electro-focusing and Isotachophoresis. Walter de Gruyier, Berlin, New York, 1977, pp. 569-564.
5 N. Good, et al., Biochemistry, 5(1966)467.

A. Adam and C. Schots (Editors), *Biochemical and Biological Applications of Isotachophoresis*   259
© 1980 Elsevier Scientific Publishing Company, Amsterdam — Printed in The Netherlands

MONITORING OF PROTEIN-PROTEIN INTERACTION BY CAPILLARY ISOTACHOPHORESIS

P. DELMOTTE

National Center for Multiple Sclerosis, Melsbroek (Belgium)

Capillary isotachophoresis of proteins is slowly gaining its place among the battery of electrophoretic techniques that can be used to analyse these compounds. For a detailed discussion of the theoretical background of this technique, the interested reader is referred to the book by Everaerts et al. (ref. 1). Technical details about protein analysis by this technique can be found in ref. 2.

By far the most interesting feature of capillary isotachophoresis resides in the fact that it provides instant qualitative and quantitative information during the separation procedure itself, without denaturation or coloration of the proteins. If real dynamic equilibrium is reached and the operating conditions are such that clearly separated peaks are obtained, then the integrated peak surfaces are a direct measure of the absolute amount of protein present in a peak (provided provision was made to assure real quantitative sample injection). Fig. 1 shows the separation patterns from 0.3 and 0.9 µl of human serum.

The straight-line relationship between amount of protein present in a peak and the integrated peak area is demonstrated in Fig. 2. The surfaces of the hatched peaks indicated in Fig. 6 were used and 0.3, 0.6 and 0.9 µl of sample was injected.

Depending on the quantitative relation between the two components, the reaction between an antigen and an antibody can result in either the formation of an insoluble complex with no antigen remaining in solution, or in the formation of soluble immune complexes, with more or less free antigen left in the solution. In the first case, detection or measurement is no problem; in the second case the problem is much more complicated. We have made some investigations in this direction and herewith describe a working model, using human albumin and

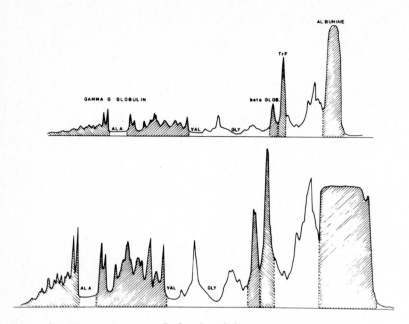

Fig. 1. Upper trace: 0.3 µl of human serum.
Lower trace: 0.9 µl of the same serum.
Hatched areas were electronically integrated.

an anti-human albumin immunoglobulin preparation (DAKO, Denmark)

A series of mixtures was made consisting of solutions with decreasing concentrations of human albumin, and of a solution containing a fixed amount of anti-albumin immunoglobulin preparation. After a suitable reaction time, the mixtures were centrifuged and the supernatants were submitted to capillary isotachophoresis. Fig 3 presents the separation patterns of one set of experiments. The upper trace is the result of injecting one of the albumin solutions mixed with an equal volume of water. As can be seen, no protein other than albumin is present. The middle trace corresponds to the anti-albumin immunoglobulin preparation. also mixed with an equal volume of water. The preparation consists for the major part of IgG but most probably also contains IgA and some other faster moving proteins. The lower trace results from the injection of the supernatant a 1:1 mixture of the two previous solutions.

It should be mentioned here that, due to the excess of antigen, the mixing of the two solutions did not result in the formation of an

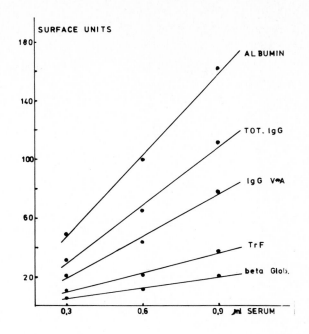

Fig. 2. Plot of integrated surfaces of peaks versus amount of protein,
for albumin, transferrin, betaglobulin, total immunoglobulin G
and the IgG fractions with mobilities between those of the
amino acids valine and beta-alanine.

insoluble immune complex and so no evidence of a reaction between
the two reactants was observed . Nevertheless, from the isotachophoretic traces it is evident that a reaction had taken place between components of the two solutions:
- The surface of the albumin peak was significantly reduced.
- Several of the immunoglobulin peaks also showed significant reduction
  in peak surface.
- Several of the fractions with mobilities between those of albumin
  and the immunoglobulin preparation showed significant augmentation
  of their peak surfaces. This must be attributed to the formation of
  soluble immune complexes.

In Fig. 4, showing other phases of this experiment, the traces of
the albumin solutions and of the supernatants from the mixture of these
solutions with equal amounts of the immunoglobulin solution are superimposed. In the lower trace, although the albumin peak is reduced in
surface, there is still much free albumin left in solution and peaks
due to soluble immune complexes can be observed. In the upper trace,

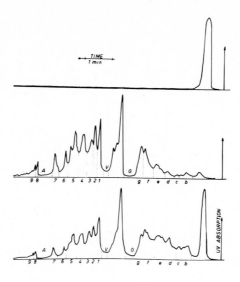

Fig. 3. Upper trace = albumin solution.
        Middle trace = anti-albumin immunoglobulin solution.
        Lower trace = supernatant of mixture.

corresponding to an albumin solution of only 2.5 mg/ml, only a very
small amount of free antigen is left in the supernatant and practically
no soluble immune complexes are evident. In all these experiments, the
peak surfaces were electronically integrated and the results are graphi-
cally presented in Fig. 5. On the horizontal axis are indicated the
final concentrations of albumin in the various mixtures with the
immunoglobulin solution. On the vertical axis of the lower figure are
indicated the percentage changes measured for the surfaces of different
fractions in the supernatants after centrifugation: the total surface
of the fractions faster than the amino acid glycine, the fraction
moving between glycine and valine, the sum of the fractions between
valine and beta-alanine and the sum of the fractions slower than beta-
alanine. Also indicated is the change in surface of fraction 9, which,
although representing only a small percentage of the total amount of
immunoglobulins, was the last fraction for which a change of surface
could still be detected (only 0.009 mg of albumin per ml was present).
In the upper part of Fig. 5 are indicated the amounts of free albumin
remaining in the supernatants. Although we cannot go into every detail

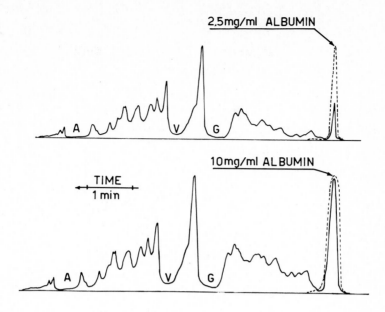

Fig. 4. Upper trace = supernatant from 1:1 mixture of 2.5 mg/ml albumin
with immunoglobulin solution.
Lower trace = idem with 10 mg albumin solution.

here, it is evident that much information can be gained from such
experiments.

In the example of a practical application, for which the traces are
shown in Fig. 6, an approach of this kind was used to detect the presence of albumin in an aqueous extract of human brain tissue. The
upper trace shows the separation pattern of the water-soluble brain
proteins. The lower trace results from the supernatant of a mixture
of the brain extract and a solution of anti-albumin immunoglobulins.
To the left side of this pattern, the peaks of brain proteins are
mixed with peaks from the immunoglobulin preparation. To the right
side , the first peak shows a significant reduction in surface (the
original peak is drawn superimposed). The albumin was spun down as
an insoluble antigen-antibody complex. This experiment not only showed
the localisation in the pattern of the albumin, but also proved that
the peak in question contained one or more other proteins (the surface
could not be further reduced by adding larger amounts of immunoglobulin).

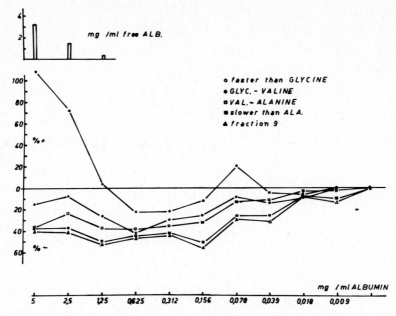

Fig. 5. Quantitative results of antigen-antibody reaction between albumin and anti-albumin immunoglobulin; for explanation see text.

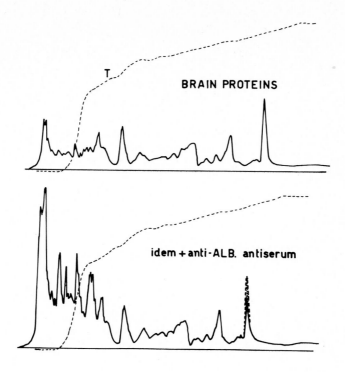

Fig. 6. Upper trace = separation pattern of soluble brain proteins.
Lower trace = separation pattern of supernatant after mixing
with anti-albumin immunoglobulin.

REFERENCES

1 F.M. Everaerts, J.L. Beckers and Th.P.E.M. Verheggen, Isotachophoresis, Elsevier, Amsterdam, 1976.
2 P. Delmotte, Sci. Tools, 24(1977)33-41.

A. Adam and C. Schots (Editors), *Biochemical and Biological Applications of Isotachophoresis*   267
© 1980 Elsevier Scientific Publishing Company, Amsterdam — Printed in The Netherlands

PROTEIN-DRUG AND PROTEIN-DETERGENT INTERACTIONS STUDIED BY CAPILLARY
ISOTACHOPHORESIS

Sven-Göran HJALMARSSON and Jörgen SJÖDAHL
LKB-produkter AB, Fack, S-161 25 Bromma (Sweden)

INTRODUCTION

Analytical isotachophoresis in a capillary tube is a`highly potent
tool for the separation of compounds exhibiting very small differences
in net electrophoretic mobility. The basic principles of isotachophoresis have been previously described by several authors (refs. 1-3)

It was anticipated that upon complex formation of a protein with
a drug or detergent it should be possible to distinguish between the
complex and the native protein using analytical isotachophoresis.
Furthermore, this method seemed feasible for separating and quantitating
the non-complexed compound.

Protein-drug interactions

The protein-drug interacting system studied in this work is composed
of the protein human serum albumin (HSA) and the anti-inflammatory drug
indomethacin [1-(p-chlorobenzoyl)-5-methoxyl-2-methyl-indole-3-acetic
acid] . This drug is frequently used for the treatment of rheumatic
diseases; its molecular formula is shown in Fig. 1.

The binding of indomethacin (I) to human plasma has been studied by
Hvidberg et al. (ref. 4), who found fifteen binding sites for the drug,
all of which were proposed to be located on the albumin molecule.
However, depending on the experimental technique used, Hultmark et al.
(ref. 5) found indomethacin to bind to four or five binding sites on
the HSA molecule.

The HSA-I system was chosen for this first approach of using analytical isotachophoresis in protein-drug complex formation studies, since
it is a relatively well studied system involving a strongly protein-
bound drug.

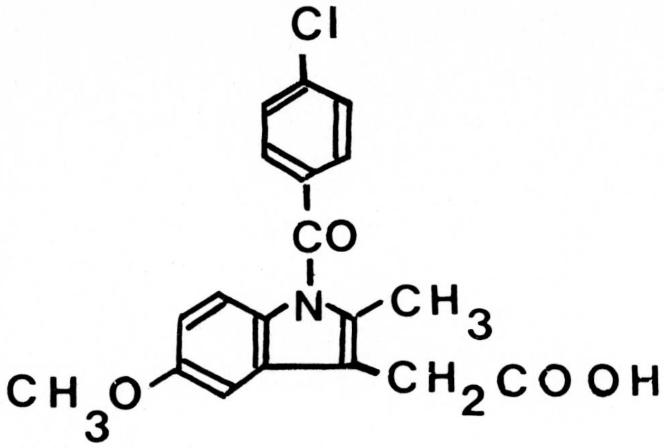

Fig. 1. Molecular formula of the drug indomethacin.

Protein-detergent interactions

In recent years considerable efforts have been put into the study of protein-detergent interactions, in order to construct a model for the natural lipid-protein interactions in membranes.

The binding of detergents to proteins involves both coulombic and hydrophobic interactions (refs. 6 and 7). The alkyl chain length of the detergent, as well as the nature of the ionic group, have been shown to influence the interactions greatly. These aspects of protein-detergent interactions have recently been reviewed by Tanford and Reynolds (ref. 8) and by Helenius and Simons (ref. 9). Detergent-protein interactions have been studied by a number of different methods, such as equilibrium dialysis (refs. 10 and 11), nuclear magnetic resonance (refs. 12 and 13), ultracentrifugation (ref. 14), intrinsic viscosity (ref. 15) and microcalorimetry (ref. 16). In most of the investigations sodium dodecyl sulphate (SDS) has been chosen as the interacting detergent and serum albumin as the protein. Depending on the method used, native serum albumin has been shown to have 8-10 high-affinity binding sites to SDS (refs. 12, 13, 17 and 18). The binding constants are very large (K = 1.0 . 10^6-1.2 . $10^6 M^{-1}$) and roughly independent of pH (ref. 7).

When the native structure is lost the binding sites are destroyed, and the binding properties of the new conformations are quite different from those of the native protein. The binding of detergent anions to the native protein can occur one molecule at a time, while the binding to the denatured protein is highly cooperative (ref. 17).

This study deals only with interactions of detergent with native bovine serum albumin (BSA) and ovalbumin (OA). These two protein were chosen because their interactions with SDS are relatively well studied, and their interactions differ considerably. Whereas native serum albumin has been shown to have 8-10 high-affinity binding sites to SDS, native ovalbumin has none (ref. 11).

MATERIALS AND METHODS

Protein-drug interactions

The human serum albumin and indomethacin were obtained from AB Kabi, Stockholm, Sweden and A.-S. Dumex, Copenhagen, Denmark, respectively. The protein was dissolved in distilled water to a concentration of 85 µg/µl (1.3 nmol/µl). Indomethacin was dissolved in the leading electrolyte (see below) to a concentration of 0.65 µg/µl (1.8 nmol/µl). Incubation mixtures were prepared from these two stock solutions according to Table 1. The samples were incubated at ambient temperature for 1 h before analysis.

The leading electrolyte was 5 mM HCl, 10 mM 2-amino-2-methyl-1,3-propanediol (ammediol, Sigma, St. Louis, Mo., U.S.A.) with 0.4 % HPMC (hydroxypropylmethylcellulose, Methocel 90 HG, 15 000 cps, Dow Chemical Co., Midland, Mich., U.S.A.), pH 9.0. Chloride acts as the leading ion and ammediol as the counter-ion; and HPMC was added to counteract electroendosmosis. The terminating electrolyte was 5 mM ε-aminocaproic acid (Roth, Karlsruhe, G.F.R.) adjusted to a pH of about 10.5 with freshly made Ba(OH)$_2$ solution (Merck, Darmstadt, G.F.R.). The separations were performed at 12°C with a constant current of initially 170 µA during detection.

Due to the high UV absorption of the protein as well as of the drug, a standard UV detector (254 nm) was used. The samples were injected in volumes of 2.5 or 5 µl.

TABLE 1

Composition of incubated samples of HSA and indomethacin

Experimental details are given in Materials and Methods

Sample (refers to Fig. 3)	Volume albumin solution (μl)	Volume indo- methacin solution (μl)	Volume buffer (μl)	Injected volume (μl)	Injected amount of Albumin (nmol)	(μg)	Indomethacin (nmol)	(μg)	Molar ratio of albumin to indome- thacin (HSA:I)
a	50	25	175	5	1.3	85	0.9	0.33	1 : 0.7
b	50	50	150	5	1.3	85	1.8	0.65	1 : 1.4
c	50	100	100	5	1.3	85	3.6	1.3	1 : 2.8
d	50	200	–	5	1.3	85	3.2	2.6	1 : 5.5

Protein-detergent interactions

Bovine serum albumin (crystallized and lyophilized) and ovalbumin (grade V, 99% pure) were obtained from Sigma, Sodium dodecyl sulphate was the specially pure grade obtained from BDH, Poole, Great Britain.

The proteins and the SDS were used without further purification. The proteins were dissolved separately in double-distilled water to a concentration of 85 μg/μl (BSA) and 58.5 μg/μl (OA). The molecular weights used for the calculations were 69 000 for bovine serum albumin and 43 000 for ovalbumin. SDS was dissolved in distilled water to a concentration of 4 mM. Incubation mixtures were prepared from the two stock solutions of proteins and the stock solution of SDS according to Table 2 and 3. The samples were incubated for approximatively 30 min before analysis.

The leading electrolyte was 5 mM HCl adjusted with 0.5 M TRIS to pH 7.80. The leading electrolyte contained 0.25% HPMC (hydroxypropyl-methylcellulose, Methocel 90 HG, 15000 cps, Dow Chemical). The termina-ting electrolyte was 10 mM glycine (Sigma) adjusted to pH 9.0 with freshly made $Ba(OH)_2$ solution (Merck).

The analyses were performed at 10°C with a constant current of 100 μA during separation and 50 μA during detection. The total analysis time was about 10 min. Samples (2 μl) were injected by means of an ordinary microlitre syringe and the separated sample zones were detected by UV transmission at 280 nm. The isotachophoretic analyses were per-formed with an LKB 2127 Tachophor (LKB-Produkter AB, Bromma, Sweden) equipped with a 23 cm capillary tube.

TABLE 2

Composition of incubated samples of BSA and SDS

Experimental details are given in Materials and Methods

Sample (refers to Fig. 6)	Volume albumin (BSA) solution (μl)	Volume SDS solution (μl)	Volume leading electrolyte (μl)	Injected volume (μl)	Injected amount of Albumin (BSA) (nmol)	(μg)	SDS (nmol)	((μg))	Molar ratio of albumin to SDS (BSA:SDS)
a	50	–	200	2	0.49	34.0	0	0	1 : 0
b	50	100	100	2	0.49	34.0	3.2	0.9	1 : 6.5
c	50	114	86	2	0.49	34.0	3.6	1.05	1 : 7.4
d	50	130	70	2	0.49	34.0	4.2	1.2	1 : 8.5
e	50	150	50	2	0.49	34.0	4.8	1.4	1 : 9.8
f	50	200	–	2	0.49	34.0	6.4	1.85	1 : 13.0

TABLE 3

Composition of incubated samples of ovalbumin and SDS

Experimental details are given in Materials and Methods

Sample refers to Fig. 7)	Volume ovalbumin (OA) solution (μl)	Volume SDS solution (μl)	Volume leading electrolyte (μl)	Injected volume (μl)	Injected amount of Ovalbumin (OA) (nmol)	(μg)	SDS (nmol)	(μg)	Molar ratio of OA to SDS (OA:SDS)
a	50	–	200	2	0.54	23.4	0	0	1 : 0
b	50	19	181	2	0.54	23.4	0.6	0.18	1 : 1.1
c	50	50	150	2	0.54	23.4	1.6	0.46	1 : 2.9
d	50	100	100	2	0.54	23.4	3.2	0.92	1 : 5.9

RESULTS AND DISCUSSION

Protein-drug interactions

Fig 2. illustrates the UV profiles obtained from analysis of the individual components of the incubations mixtures. It is evident that a very pure preparation of HSA was used, as indicated by the homogeneity and the very sharp zone boundaries (Fig. 2a). Furthermore, indomethacin with its strongly UV-absorbing chromophore appears as a homogeneous and well-defined UV profile (Fig. 2b.). The small UV-absorbing spikes, as well as the shoulder on the indomethacin peak, were also observed when analysing a blank. Consequently, they were ascribed to impurities in the electrolyte system. From this figure, it can be concluded that

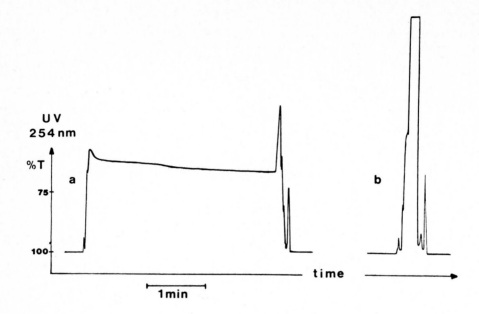

Fig. 2. UV profiles from the isotachophoretic analysis of albumin (a)
and indomethacin (b). The injected amounts were 210 µg and
0.81 µg of albumin and indomethacin, respectively.

indomethacin gives a higher UV absorption than albumin at isotachophoretic concentrations under the experimental conditions used.

The titration of the binding sites on HSA with indomethacin is illustrated in Fig. 3. It should be stressed that in isotachophoretic
analysis, the zone width and not the zone height reflects the amount
of material in each zone. Comparing the zones mutually and to the zone
of free albumin in Fig. 2a (1.3 and 3.3 nmol, respectively, were injected), it is evident that the relative zone width changes during the
titration. This may reflect a change in net electrophoretic mobility
when the drug binds to the albumin molecule. However, the significant
feature of these UV profiles is the increase in peak height when increasing the amount of indomethacin. This reflects the difference
between the UV absorption of HSA alone and that of the complex of
HSA-I_n, where n indicates the relative molar amount of bound indomethacin. Furthermore, it is evident that no or only very small amounts
of free indomethacin are present in the first three samples.It is

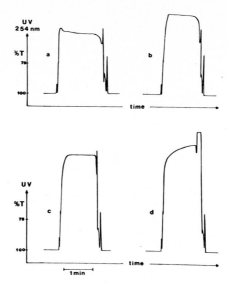

Fig. 3. UV profiles from the isotachophoretic analysis of incubation
mixtures containing various ratios of albumin to indomethacin.
Ratios of albumin to indomethacin: 1:0.7 (a), 1:1.4 (b),
1:2.8 (c), and 1:5.5 (d).

concluded that, for these molar ratios, the binding of the drug to the
protein is practically stoichiometric, i.e. n equals 0.7, 1.4 and 2.8,
respectively. However, when reaching a molar ratio of 1:5.5 (HSA:I) a
sharp indomethacin zone appears, indicating that the albumin is satured
with bound indomethacin (Fig. 3d). By measuring the width of the indomethacin zone, the molar amount can be calculated from a pre-constructed
calibration curve. In this case, the zone width corresponds to 2.2 nmol,
of indomethacin. Since the total amount of indomethacin added was
7.2 nmol (Table 1), the amount of bound indomethacin is 5.0 nmol. The
saturation ratio can now be calculated to be 1.3:5.0 which equals 1:3.9,
indicating the binding of four indomethacin molecules to each molecule
of albumin. However, the curved UV profile of the saturated albumin complex (Fig. 3d) may indicate a lower binding strength for the fourth
indomethacin molecule.

Protein-detergent interactions

Fig . 4 illustrates the isotachophoretic analysis of the electrolytes without sample (Fig. 4a) and with 4 nmol SDS added (Fig. 4b).

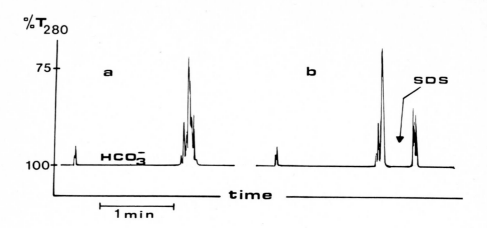

Fig. 4. UV profiles from the isotachophoretic analysis of the electrolytes without samples (a) and with 4 nmol SDS added (b).

A number of absorbing zones can be seen in both UV profiles. In
analysis 4a no sample was added, and consequently all the UV- and non
UV-absorbing parts originate from impurities in the electrolytes. The
non-UV-absorbing part originates from carbon dioxide dissolved in the
electrolyte and at the given pH it will show up as hydrogen carbonate
(HCO_3^-). The HCO_3^- zone is rather short when freshly prepared electrolytes are used and will not influence the resolution.

In the experiment illustrated in Fig. 4b, SDS has been added. Since
SDS is negatively charged at the pH used (pH = 7.8 in leading electrolyte), and has a net mobility between the net mobilities of the leading
and terminating ions, it will move as a sharply separated zone. SDS
does not absorb at 280 nm, the walvelength used here. However, SDS is
clearly distinguished on the UV profile since it acts as a non-UV-
absorbing spacer in between UV-absorbing impurities.

According to the principle of isotachophoresis, the zone width is
directly proportional to the molar amount of the sample. This is
evident from Fig. 5, which is a calibration curve for SDS.

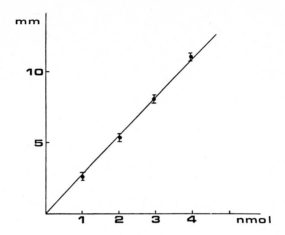

Fig. 5. Calibration curve showing the linear relationship between nmol of SDS injected and the corresponding zone width in mm measured from the UV-profile.

Bovine serum albumin (BSA): the titration of the binding sites on BSA with SDS is illustrated in Fig. 6. Fig. 6a shows the isotachophoretic analysis of 34 µg of BSA injected in a volume of 2 µl (see Table 2). It is evident that the BSA is not homogeneous, since a sharp and re-producible zone occurs prior to a long homogeneous zone. However, the front peak does not influence the titration results, since it obviously remains almost unchanged throughout the titration.

When SDS is added in a molar ratio of 1:6.5 (BSA:SDS), it is evident that no free SDS is present in the sample (Fig. 6b). However, when a molar ratio of 1:7.4 (BSA:SDS) is reached, a narrow zone of SDS appears, indicating that the albumin is satured with bound SDS (Fig. 6c). From these three experiments it could already be concluded that each BSA molecule binds seven SDS molecules. This binding number can, however, also be verified by adding more SDS to each BSA molecule and calculating the amount of free SDS from the calibration curve prepared previously. The zone width of the free SDS in Fig. 6f corresponds to 2.9 nmol SDS. Since the total amount of SDS added was 6.4 nmol (Table 2), the amount of bound SDS is 3.5 nmol. The saturation ratio can now be calculated to

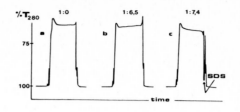

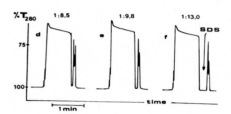

Fig. 6. UV profiles from the isotachophoretic analysis of incubation
mixtures containing various molar ratios of bovine serum albumin
to SDS. Ratios of albumin to SDS: 1:0 (a), 1:6.5 (b), 1:7.4 (c),
1:8.5 (d), 1:9.8 (e), and 1:13.0 (f).

be 0.49:3.5 which equals 1:7.1 and indicates the binding of seven SDS
molecules to each molecule of BSA. The same result was found when
running the incubated sample in a longer capillary tube (43 cm), indicating that the isotachophoretic equilibrium was fully reached and that
no continuous disruption of the complex was occuring. The number of
high-affinity binding sites of BSA for SDS found by isotachophoresis
is in good agreement with those results found by other more elaborate
techniques. The value obtained by techniques such as equilibrium dialysis and nuclear magnetic resonance is often found to be 8-9 (refs. 7,
12 and 13). Since the experimental errors are in general about 10%, the
value "7" obtained in this study is in good agreement with the earlier
investigations.

The over all accuracy of the values obtained by isotachophoresis is
found to be within 2-3%, the main errors being in preparation (weighing
and dissolving) of the solutions, injection of sample and measurement
of zone length.

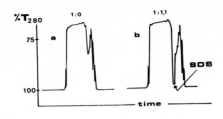

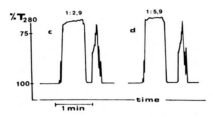

Fig . 7. UV profiles from the isotachophoretic analysis of incubation
mixtures containing various molar ratios of ovalbumin to SDS.
Ratios of ovalbumin to SDS: 1:0 (a), 1:1.1 (b), 1:2.9 (c) and
1:5.9 (d).

Ovalbumin (OA) : Ovalbumin in its native state lacks high-affinity
binding sites for SDS (ref. 11). The OA was titrated with SDS according
to Table 3. Fig 7b illustrates the UV profile of the incubated sample
of OA and SDS in the molar ratio 1:1.1. It is evident that a zone of
free SDS occurs, and from the corresponding zone length it is obvious
that no SDS is bound to the OA. When the molar ratio of SDS to OA was
increased (Fig. 7d), the zone width of SDS was also increased and no
high-affinity interactions could be found. This result was in complete
agreement with the results found by equilibrium dialysis (ref. 11).

It is concluded that analytical isotachophoresis in a capillary
tube, due to the very high resolution of the technique in combination
with the possibility of quantitation, is a powerful tool for studying
strong complex-formation systems, such as the protein-drug and protein-
detergent interactions studied in this work.

REFERENCES

1 H. Haglund, Sci. Tools, 17, N° 1(1970)2-12.
2 L. Arlinger, J. Chromatogr., 91(1974)785-794.
3 F.M. Everaerts, J.L. Beckers and Th;P.E.M. Verheggen, Isotachophoresis - Theory, Instrumentation and Applications (Journal of Chromatography Library, Vol. 6), Elsevier, Amsterdam, 1976.
4 E. Hvidberg, H.H. Lausen and J.A. Jansen, Eur. J. Clin. Pharmacol, 4(1972)119-124.
5 D. Hultmark, K.G. Borg, R. Elofsson and L. Palmer, Acta Pharm. Suec., 12(1975)259-276.
6 J. Steinhardt and J.A. Reynolds, Multiple Equilibrium in Proteins, Academic Press, New York, 1969, p. 234.
7 J.A. Reynolds, J.P. Gallagher and J. Steinhardt, Biochemistry,9(1970) 1232.
8 C. Tanford and J.A. Reynolds, Biochim. Biophys. Acta, 457(1976)133-170.
9 A. Helenius and K. Simons, Biochim. Biophys. Acta, 415(1975)29.79.
10 J.A. Reynolds and C. Tanford, Proc. Nat. Acad. Sci. U.S., 66(1970) 1002-1007.
11 J. Steinhardt, N. Stocker, D. Carroll, and K.S. Birdi, Biochemistry, 21(1974)4461-4468.
12 G. Gillberg-La Force and S. Frosén, Biochem. Biophys. Res. Comm., 38(1970)137-142.
13 J.-E. Norne, S.-G. Hjalmarsson, B. Lindman and M. Zeppezauer, Biochemistry, 14(1975)3401-3408 .
14 S. Clark, J. Biol. Chem., 250(1975)5459-5469 .
15 J.A. Reynolds and C. Tanford, J. Biol. Chem., 245(1970) 5161-5165.
16 K. Kale, G.C. Krescheck and G. Vanderkooi, Biochim. Biophys. Acta, 535(1978)334-341.
17 C. Tanford, Advan. Protein Chem., 23(1968)211-215.
18 J.A. Reynolds, S. Herbert, H. Polet and J. Steinhardt, Biochemistry, 3(1967)937-947.